BIOCHEMICAL ASPECTS OF NEUROLOGICAL DISORDERS

THIRD SERIES

BIOCHEMICAL ASPECTS OF NEUROLOGICAL DISORDERS

THIRD SERIES

EDITED BY

JOHN N. CUMINGS

M.D. F.R.C.P. F.C.Path.

Professor of Chemical Pathology
Institute of Neurology, National Hospital, London

AND

MICHAEL KREMER

M.D. B.Sc. F.R.C.P.

Consultant Neurologist
National Hospital and Middlesex Hospital, London

BLACKWELL SCIENTIFIC PUBLICATIONS
OXFORD AND EDINBURGH

LR J EFEL

SBN 632 04970 7

FIRST PUBLISHED 1968

Printed in Great Britain by Alden & Mowbray Ltd
at the Alden Press, Oxford
and bound at Kemp Hall Bindery

CONTENTS

LIST OF CONTRIBUTORS

P. N. CAMPBELL Ph.D., D.Sc.
Professor of Biochemistry
University of Leeds

R. E. CASPARY M.Sc.
on Scientific Staff, Medical Research Council in
Demyelinating Diseases Research Unit, Newcastle upon Tyne

SYDNEY COHEN M.D., Ph.D., F.C.Path.
Professor of Chemical Pathology
Guy's Hospital Medical School, London SE1

JOHN N. CUMINGS M.D., F.R.C.P., F.C.Path.
Professor of Chemical Pathology
Institute of Neurology, National Hospital, Queen Square, London WC1

G. CURZON Ph.D.
Senior Lecturer in Chemical Pathology
Institute of Neurology, National Hospital, Queen Square, London WC1

HUGH DAVSON D.Sc.
Fellow and Research Associate
on Medical Research Council External Staff
University College, London WC1

A. M. JELLIFFE M.D., D.C.H., F.F.R.
Consultant Radiotherapist
Middlesex Hospital, London W1

MICHAEL KREMER M.D., B.Sc., F.R.C.P.
Consultant Neurologist
National Hospital, Queen Square, London WC1
Middlesex Hospital, London W1

P. T. LASCELLES M.D., M.C.Path.
Senior Lecturer in Chemical Pathology
Institute of Neurology, National Hospital, Queen Square, London WC1

W. A. LISHMAN M.D., M.R.C.P., D.P.M.
Consultant Psychiatrist
Bethlem Royal Hospital, Beckingham, Kent
Maudsley Hospital, Denmark Hill, London SE5

VALENTINE LOGUE F.R.C.S.*
Consultant Neurosurgeon
The National Hospitals for Nervous Diseases, London WC1
Middlesex Hospital, London W1

JOHN MARSHALL M.D., F.R.C.P.(Ed.), F.R.C.P., D.P.M.
Reader in Clinical Neurology
Institute of Neurology, National Hospital, Queen Square, London WC1

J. R. A. MITCHELL M.A., M.D., M.R.C.P., D.Phil.
Professor of Medicine
University of Nottingham

R. RITCHIE RUSSELL C.B.E., M.D., D.Sc., F.R.C.P.
Professor of Neurology
United Oxford Hospitals

JOHN WILSON M.B., M.R.C.P., Ph.D.
Consultant Neurologist
Hospital for Sick Children, Great Ormond Street, London WC1

O. H. WOLFF M.D., F.R.C.P.
Professor of Child Health
Institute of Child Health, Hospital for Sick Children
Great Ormond Street, London WC1

L. I. WOOLF Ph.D.†
on Medical Research Council External Staff
Department of the Regius Professor of Medicine, Radcliffe Infirmary, Oxford

K. J. ZILKHA M.D., M.R.C.P.
Consultant Neurologist
National Hospital, Queen Square, London WC1
King's College Hospital, Denmark Hill, London SE5

* Now Professor of Neurosurgery, Institute of Neurology.
† Present address: Division of Neurological Sciences, Department of Psychiatry, University of British Columbia, Vancouver 8, Canada.

PREFACE

The third series of lectures in this book is again based on those given at the Institute of Neurology of the University of London at the National Hospital, Queen Square. The lectures were all delivered in the Autumn Term of 1967. On this occasion they were sponsored by the Sandoz Foundation for advanced lectures to whom we owe our very grateful thanks.

Though the lectures appear under the same general title as the previous volumes, the contents are very different. As will be seen from the choice of subjects we have interpreted both the clinical and the biochemical aspects very widely indeed. The lecturers were given considerable lattitude in the coverage of their topics. This policy was deliberate not only to avoid repetition of the previous series but to make the lectures of interest to a less limited medical audience.

We have retained the same general pattern as previously, the lectures being given in pairs. The first deals mainly with clinical features whereas the second covers the laboratory research side, thus giving a reasonable contrast in approach.

The lectures are, as before, presented almost exactly as delivered and serve therefore as an introduction to the subjects rather than a complete survey, though a list of references is given after each chapter, prepared by each author who is solely responsible for their accuracy.

We must thank the lecturers who responded so readily to our request and our publishers who have given continual help. We deeply appreciate and offer our thanks to our secretarial and technical staff for their assistance during the lectures and their help in the final preparation of this volume.

J.N.C.

M.K.

CHAPTER ONE

SOME NEUROLOGICAL DISORDERS IN RELATION TO AUTOIMMUNITY

MICHAEL KREMER

The changes which result from any type of lesion in the peripheral nervous system can only be understood after a careful study of the normal structure. McDonald (1967) gives the following description. Each nerve fibre consists of an axon arising from the nerve cell body and the Schwann cell in which the axon is embedded. Myelin is formed by the spiral wrapping of an invagination of the Schwann cell surface membrane, the mesaxon, around the axis cylinder. The adjacent layers of the spiral will fuse, but a thin layer of cytoplasm is found inside and outside the layer of myelin. In the unmyelinated fibres several axons are invaginated in each Schwann cell but the mesaxons remain short and simple. In myelinated fibres, the Schwann cell nucleus and related organ cells are found in a bulge of undifferentiated cytoplasm in the outer cytoplasmic layer in the middle of the cell. Around the outer surface of the Schwann cell is a prominent basement membrane. The myelin sheath is regularly interrupted by the nodes of Ranvier and each internodal segment is formed by one Schwann cell and contains a variable number of the slender incisuras of Schmidt-Lauterman, formed by separation of the myelin lamellae. In the largest fibres the internodal length may be 1 mm or more.

The blood supply of the peripheral nerves is rich, consisting of finely meshed plexuses of capillaries, arterioles and venules distributed longitudinally in the perineurium within the central parts of the nerves and between the axons. Sunderland (1945) has described great variations in the vascular pattern in the peripheral nerves; the epineural anastomosis, the intrinsic longitudinal plexus, and the free anastomosis that these provide between adjacent segmental vasa nervorum usually ensures abundant compensatory circulation. Careful painstaking work on the peripheral nerves, with study in particular of many sections both longitudinal and transverse and the study also of individual nerve fibres by teasing techniques, have shown that the two major pathological processes are degeneration by damage to the parent cell involving the dying back process described by Cavanagh (1964), and demyelination, which is a segmental process with the Schwann cell as the sufferer and preservation of the axons, the Gombault (1880) phenomenon. In some conditions both processes can exist together, but this varies in different species exposed to the same traumatic insult.

From the clinical angle it has been shown that electrical conduction ceases in degenerating nerve when the axons are destroyed, but at earlier stages the conduction velocity decreases by 10–20 per cent. In demyelination, conduction is delayed by local changes and over nerve trunks by 60 per cent or more. Similar delay is seen also in remyelination. Perhaps some of this delay is due to block occurring in the fastest fibres. The diagnostic point made by Gilliatt (1966) is that if the conduction delay is more than 40 per cent then demyelination is the likely pathological process.

EXPERIMENTAL ALLERGIC NEURITIS
Although the lesions of experimental allergic encephalomyelitis affect predominantly the central nervous system, they occur also in peripheral nervous tissue. However, Waksman and Adams (1955, 1956) demonstrated that injection of rabbits and mice with homogenates of homologous or heterologous peripheral nerve tissue combined with Freund's adjuvant resulted in the development of lesions resembling those of experimental allergic encephalomyelitis but confined to the peripheral nervous system. This condition was called experimental allergic neuritis and it was concluded that there were antigenic differences in the myelin of peripheral and central nervous tissue. Injection of guinea-pigs with peripheral nervous tissue and adjuvant resulted in experimental allergic neuritis in some animals while others developed lesions of both central and peripheral nervous tissues. Experimental encephalomyelitis is described in another chapter but this may be anticipated briefly by saying that allergic neuritis may show similar phenomena. These are the latent period before its onset, the acute course, and the development of lesions consisting of perivascular infiltration of nervous tissues with histiocytes, lymphocytes, eosinophils and plasma cells associated with demyelination without destruction of axon cylinders. The disease was also shown to resemble experimental allergic encephalomyelitis in that it was not prevented by autoclaving the nervous tissue incorporated with the inducing inoculum. Tests for antibody and for delayed hypersensitivity were positive in rabbits and mice

following the injection of peripheral nervous tissue preparations and adjuvant, but the occurrence of neuritis was very irregular and it was concluded that multiple antigen–antibody systems were involved in the reaction and that until all the antigenic factors responsible could be purified the significance of positive tests could not be determined. Passive induction of experimental allergic neuritis has been produced in rabbits by transfer of lymph node cells from affected rabbits and the conditions are the same as for passive induction of experimental allergic encephalomyelitis. The distribution of the lesions of this experimental neuritis varies in different species. They are mainly in the spinal ganglia, spinal roots and peripheral nerves.

Two possible immunological mechanisms are postulated in both experimental allergic encephalomyelitis and experimental neuritis.

1 Antibodies to nervous tissue antigens gain access to, and react with, nervous tissue. The damage then possibly results from the additional effect of complement.

2 Damage results from the development and migration of sensitized lymphoid cells to the central nervous system, that is delayed hypersensitivity reaction. In both cases myelin is the source of the encephalitic and neuritic factor.

The pathological features and course of experimental allergic neuritis are sufficiently like the Guillain-Barré syndrome to raise the possibility of an autoimmune pathogenesis for this condition. There is not much direct support for this, but Melnick (1963) described tests for antibodies on the sera of thirty-eight patients. Using saline extracts of nervous tissues as antigens, negative results were obtained in precipitin and tanned red cell agglutination tests but complement-fixation tests were positive at serum dilutions greater than 1:8 in 50 per cent. Some of the sera reacted with extracts of various other tissues, but the titres were usually higher with nervous tissue. Positive results were obtained with 31 per cent of sera from patients with hypersensitivity diseases (mostly systemic lupus erythematosus and cirrhosis), but these sera reacted more strongly with extracts of other tissues than did the Guillain-Barré sera and presumably contained autoimmune complement-fixation antibodies. Sera from patients with other neurological disorders, including other demyelinating diseases, gave very few positive results. It was felt that these results were consistent with the development of a complement-fixing serum factor, possibly antibody, which reacted specifically with nervous tissue. In a group of patients with the Guillain-Barré syndrome the demonstration of high complement-fixation titres in sera from two patients on the second day of the illness suggested that the serum factor appears before the onset of clinical neurological changes. It must be added that in this investigation positive sera reacted with central as well as peripheral nervous tissue extracts. This is similar to the findings in complement-fixation tests in rabbits with experimental allergic neuritis. In both the Guillain-Barré syndrome and experimental allergic neuritis it seems unlikely that the antigens in the complement-fixation tests are the same as the peripheral nervous agent which induces experimental allergic neuritis, but tests cannot be regarded as firm evidence that the lesions in the Guillain-Barré syndrome have an autoimmune pathogenesis, but the histological findings are sufficiently similar to make this more than a possibility.

THE RESPONSE OF NERVOUS TISSUE IN
THE COLLAGEN DISORDERS

The problem here is in the definition of the collagen or connective tissue diseases. The so-called collagen disorders or connective tissue diseases are a group of conditions, recognized easily when typical, which frequently shade into one another and in which immunological tests give varying and frequently overlapping responses. This means that the criteria for diagnosis are not precise and vary in different centres, and that so-called characteristic clinical features of one 'entity' are often found to have biopsy findings which are uncertain or point to one of the other members of this diffuse group.

The involvement of the nervous system in these disorders is by two main pathological processes. The first is that produced by vascular involvement and is therefore best seen in polyarteritis nodosa. This disease is characterized by foci of degeneration in or near the branching or bifurcation of medium-sized or small arteries. The acute lesions show oedema of the media and perivascular tissues followed by infiltration with polymorphonuclear and eosinophil leucocytes. Later there is a fibrinoid degeneration in the vessel wall, or necrosis. In the healing process fibrosis is most dense in the adventitia but may extend through the whole thickness of the vessel wall. The end-result is either occlusion or weakening of the wall, with aneurysm formation and varying stages of destruction, but repair may also be found. For years the popular hypothesis has been that polyarteritis nodosa is a 'hypersensitivity disease'. The vascular lesions in this condition are said to resemble those in serum sickness and in hypersensitivity to drugs such as penicillin, sulphonamides and iodides. Similarly, vascular lesions such as these can be produced experimentally in animals by sensitization to foreign proteins. In experimental serum sickness there is evidence that a soluble antigen–antibody complex is deposited in the tissues, but there is no direct evidence for such a mechanism in polyarteritis nodosa. Rheumatoid factor is found to be positive in 15 per cent and nuclear antibodies present in a few, but these may be intermediate forms close to rheumatoid arthritis or systemic lupus erythematosus as the majority of patients with polyarteritis nodosa have no serum autoantibodies nor evidence of drug or other allergies and the evidence that autoimmunity may play a part in the pathogenesis of polyarteritis nodosa is very thin indeed.

Be that as it may, the neurological lesions are frequently dependent on ischaemia. The presence of asymmetrical polyneuritis, the so-called mononeuritis multiplex, is well known, and biopsy of the nerve and vaso nervorum as described by Blehen, Lovelace, and Cotton (1963) may be necessary to confirm the diagnosis. The interesting clinical features are not only the haphazard distribution but the rapid fluctuations in function in the nerves affected. The rich blood supply and the changes in the vasa nervorum are responsible for functional blocks and rapidly reversible focal or segmental demyelination which enables recovery of some function to occur more rapidly than would be thought possible. An interesting group of peripheral lesions were described by Pallis and Scott (1965) in rheumatoid arthritis with vascular occlusion. Ischaemic changes in the brain and spinal cord can obviously occur and need no comment.

The second mechanism of neurological involvement in the collagen diseases is more speculative and consists of generalized demyelinating neuropathies. It is important that these should not be confused with the very diffuse or coalesced mononeuritis multiplex. Here both peripheral and central neuropathies occur and the mechanism is unknown. It is very tempting to include some of the carcinomatous neuropathies in this group. Croft, Henson, Urich and Wilkinson (1965) described four cases of sensory neuropathy in which all sensory modalities were lost with pseudoathetosis and ataxia, frequent elevation of cerebrospinal fluid protein, and a circulating organ-specific antibody to brain in the sera of all four patients and in the cerebrospinal fluid of two. The morbid anatomy showed extensive destruction of posterior root ganglia and mild changes in parts of the brain. The authors postulate either a possible primary immunological insult or possible virus disorder or both. In some respects this is similar to experimental allergic neuritis. Whether the changes seen in other carcinomatous neuropathies are related is as yet impossible to say, but much work on antigen–antibody and autoantibody identification must precede this.

Neuropathies are found with macro-globulinaemia and in a case reported by Dayan and Lewis (1966) a macro-cryo-globulinaemia was implicated. These cases show a greater or lesser degree of demyelination and similar cases are found in myelomatosis, lymphosarcoma, and Hodgkin's disease. These conditions may be associated with an altered immunological state but, as has been seen in other conditions, this does not of itself establish an autoimmune hypothesis.

MYASTHENIA GRAVIS

The mechanism of myasthenia gravis is at present still varying between that of an autoimmune disease and some other possibilities. The evidence must be examined in a little detail to see how difficult the problem really is. Histological examination of the muscles usually show lymphocytic aggregations, the lymphorrhages, in relation to focal atrophy and sarcoplasmic basophilia of individual muscle fibres. Necrosis of fibres with an inflammatory reaction and atrophy of fibres without loss of striation or inflammation are also seen and even the myocardium may be involved. The structural changes are found most readily in clinically affected muscles, but they are often rare and do not appear to have accounted for the functional changes, nor are they specific for myasthenia gravis. Lymphorrhages have been found in some cases of rheumatoid arthritis and the other connective tissue disorders and in Addison's disease. The other changes can be seen in some myopathies and polymyositis. Physiological and pharmacological studies suggest a block in the neuromuscular end organ as the cause of this abnormality of function. Structural abnormality in the end organs have been demonstrated both by light and electron microscopy and it is not known whether they are the cause or result of the block or whether they are specific for myasthenia gravis. Simpson (1960, 1963) suggested that the disorder might result from autoimmunization and postulated blockage by an autoantibody of the muscle receptor for acetylcholine. Nastuk, Plescia and Osserman (1960) previously reported abnormal variations in the level of serum complement in myasthenia gravis, the level being low especially in exacerbations of the disease. This suggested the possibility of an antigen–antibody reaction taking place *in vivo* with consequent utilization of complement. Direct evidence of the occurrence of antibody in skeletal muscle was shown by Strauss, Seegal, Hsu, Buckhold, Nastuk and Osserman (1960) who reported the detection of a serum globulin factor which reacted in direct immunofluorescence tests with skeletal muscle fibres. Since then there have been reports of antibody to skeletal muscle using immunofluorescent technique, antiglobulin consumption, complement fixation, and tanned red cell agglutination. Antibody is not demonstrable in all cases of myasthenia gravis, 42 per cent have been reported in a large series by Feltkamp, van der Geld, Kruiff and Oosterhuis (1963) and 52 per cent in seventy-four patients reported by Downes, Greenwood and Wray (1966) from the National Hospital and the Middlesex Hospital. Further testing has suggested that the muscle antibody reacts also with thymus but not with lymph node, indicating a single antibody. Although transfusion of large volumes of blood from myasthenic patients to a healthy volunteer did not produce any myasthenic symptoms, it is well known that newborn infants of myasthenic mothers may present features of myasthenia gravis, which disappear within a few weeks of birth and do not recur. This suggests transplacental transfer of a humeral factor responsible for the muscular weakness and raises the possibility that antibody to muscle may play some part. It is known that 7S antibodies may cross the placenta and that they are metabolized in the first few weeks of life, during which time the myasthenic baby recovers spontaneously. The transplacental passage of muscle antibodies has not yet been demonstrated and neonatal myasthenia has been described in the child of a mother in whom no circulating antibodies were found. Sticker, Tholen, Massini and Staub (1960) reported that cases of myasthenia gravis are temporarily benefited by haemodialysis, this effect gradually disappearing after 1–2 weeks. This was not due to changes in potassium and suggests the presence of a pharmacologically active substance capable of crossing a dialysis membrane and of much lower molecular weight than any antibody. While it seems unlikely that compounds of low molecular weight could produce effects for several weeks, the neuromuscular end organs of normal infants have been shown to have several behavioural features resembling, though less marked than, those in myasthenia gravis and it may be that the infant is particularly susceptible to the postulated humeral blocking agent.

The morphological changes in the muscles in myasthenia gravis, especially the lymphorrhages, could be interpreted as indicative of an autoimmune reaction of the delayed hypersensitivity type, but cellular infiltration of the neuromuscular end organs is not a feature of the disease and there is no evidence to suggest that the functional disturbance is attributable to a delayed hypersensitivity reaction change in the thymus.

The recent realization that the thymus plays an important role in the development of lymphoid tissue and immunological responsiveness has almost coincided with the demonstrations of autoantibodies to skeletal muscle in myasthenia gravis, and the thymic abnormalities commonly present in myasthenia are therefore of great interest. In approximately

70 per cent of cases, lymphoid follicles with active germinal centres are present in the medulla, and 15 per cent have a thymic tumour usually of epithelial cell type and often non-malignant. Cases with thymomas often show germinal centres in the residual thymic tissues. Now it has been established that lymphoid germinal centres are indicative of an immuno-logical response and they are absent or scanty in normal thymic tissue and it does seem likely that the thymus is not a normal site of production of antibody or sensitized lymphoid cells. There is some evidence that thymic cells injected into histo-incompatible animals, mainly capable of reacting immunologically against the host, and injection of antigen into the thymus is followed by the development of germinal centres and plasma cells which contain corresponding antibodies. It has been demonstrated that the germinal centres in the thymic medulla of patients with myasthenia gravis contain immunoglobulin G and it is thus probable that antibodies are being produced, but it is not known which is the antigen producing this response. It has already been mentioned that some myasthenic sera react with muscle fibres and with thymic cells, thus supporting the possibility that the germinal centres of the thymus represent an autoimmune thymitis comparable to chronic thyroiditis.

It is interesting that myasthenia gravis associated with a thymoma tends to be severe and that such cases have higher incidences and titres of anti-bodies to skeletal muscle than those without thymic tumour.

The results of thymectomy in myasthenia gravis are irregular. The problem is complicated in any series by the variation of surgical techniques and occasionally by uncertainty about complete clearance of thymic tissues, though this is much less likely now. Sometimes there is an immedi-ate response giving disturbing reactions to hitherto well-tolerated doses of anti-cholinesterase substances. This may sometimes lead to mistaken identification of cholinergic and myasthenic crises (Simpson, 1963). Sometimes the response is delayed for months or even 1–2 years, but whether this is related to a progressive drop in the circulating muscle antibodies or complement fixation is uncertain as the test results are incon-clusive. What, then, must be the criteria for recommending thymectomy. Personal experience based on patients operated on by two surgeons of great skill has shown that the real indication for thymectomy is fairly generalized myasthenia and evidence of progression of the disease as shown by the need for increasing doses of anticholinesterases. While it is true that this is more likely in early acute cases, the change from the stationary mild form to the progressive more severe form may occur at any time and then calls for thymectomy. The presence of a thymoma is not necessarily a contraindication when good team work for post-operative management is available.

The association of myasthenia with other diseases of the so-called autoimmune type is well described, but it is not yet possible to say more than that myasthenia gravis has features which are compatible with the autoimmune hypothesis, but that no firm or direct evidence for an absolute link is yet available. Autoimmunity is probably only part of the story.

REFERENCES

BLEHEN S.S., LOVELACE R.E. and COTTON R.E. (1963) Q. *Jl Med.* **32**, 193

CAVANAGH J.B. (1964) *J. Path. Bact.* **87**, 365

CROFT P.B., HENSON R.A., URICH H. and WILKINSON P.C. (1965) *Brain* **88**, 501

DAYAN A.D. and LEWIS P.D. (1966) *Neurology (Minneap.)* **16**, 1141

DOWNES J.M., GREENWOOD B.M. and WRAY S.H. (1966) Q. *Jl Med.* **35**, 85

FELTRAMP T.E.W., VAN DER GELD H., KRUIFF K. and OOSTERHUIS H.J.G.H. (1963) *Lancet* **i**, 667

GILLIATT R.W. (1966) *J. Roy. Coll. Phycns (Lond.)* **1**, 50

GOMBAULT M. (1880) *Archs Neurol. (Paris)* **1**, 11

McDONALD I.W. (1967) In *Modern Trends in Neurology*, p. 145. Ed. WILLIAMS D. Butterworth, London

MELNICK S.C. (1963) *Brit. med. J.* **1**, 368

NASTUR W.L., PLESCIA O.J. and OSSERMAN K.E. (1960) *Proc. Soc. exp. Biol. Med.* **108**, 177

PALLIS C.A. and SCOTT J.T. (1965) *Brit. med. J.* **1**, 1141

SIMPSON J.A. (1960) *Scott. med. J.* **5**, 419

SIMPSON J.A. (1963) In *Biochemical Aspects of Neurological Disorders*, 2nd ed., p. 53. Ed. CUMINGS J.N. and KREMER M. Blackwell Scientific Publications, Oxford

STRAUSS A.J.L., SEEGAL B.C., HSU K.G., BURKHOLDER P.M., NASTUR W.L. and OSSERMAN K.E. (1960) *Proc. Soc. exp. Biol. Med.* **105**, 184

STRICKER E., THOLEN H., MASSINI M.A. and STAUB H. (1960) *J. Neurol. Neurosurg. Psychiat.* **23**, 291

SUNDERLAND S. (1945) *Archs Neurol. Psychiatry* **53**, 91

WAKSMAN B.H. and ADAMS R.D. (1955) *J. exp. Med.* **102**, 213

WAKSMAN B.H. and ADAMS R.D. (1956) *J. Neuropath. exp. Neurol.* **15**, 293

ADDITIONAL READING

ANDERSON J.R., BUCHANAN W.W. and GOUDIE R.B. (1967) In *Autoimmunity*, Charles C. Thomas, Springfield, Ill., U.S.A.

CHAPTER TWO

THE IMMUNE RESPONSE IN RELATION TO THE NERVOUS SYSTEM

SYDNEY COHEN

Acquired immunity depends essentially upon a process of lymphoid cell proliferation which occurs in response to antigenic stimulation and leads to the production of serum immunoglobulins and probably also sensitized cells having a specific affinity for the antigen. Both the afferent and efferent components of this response are dependent upon vascular and lymphatic connections between the antigen site and the immune system allowing for the passage between them of lymphoid cells and macromolecules. It is sometimes suggested that the specialized vascular architecture of the nervous system may modify its interaction with the immunological system. The evidence in regard to this problem is considered in this paper in general terms.

THE IMMUNE RESPONSE TO CIRCULATING ANTIGENS (Fig. 1)
Several *in vitro* and *in vivo* experiments have indicated that the initial step in the immune response to soluble and particulate antigens involves their phagocytosis by macrophage cells (see Mitchison, 1968). This is shown, for example, by the fact that antigen which has circulated in one animal for some hours so that all phagocytosable material is removed is no longer immunogenic when tested in other recipients (Frei, Benacerraf and Thorbecke, 1965). Indeed, such material may induce a long-lasting state of specific immunological paralysis, possibly by direct interaction with lymphoid cells of appropriate specificity (Fig. 1). The way in which antigen ingested by macrophages is able to stimulate lymphocytes to produce antibody remains unknown. There is evidence of close physical contact between lymphocytes and macrophages, and highly antigenic material, sometimes associated with RNA, has been extracted from macrophages after immunization (see Braun and Cohen, 1968). Material liberated from macrophages must interact with lymphocytes, perhaps selectively with

those which carry antibody of the appropriate specificity on the cell surface. The stimulated lymphocytes undergo repeated mitotic divisions and differentiate into larger cells, some of which resemble plasma cells, and contain and secrete specific antibody.

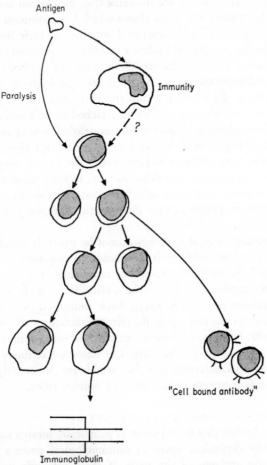

FIG. 1. Diagrammatic representation of the immune response. Phagocytic cells ingest antigen and liberate material which stimulates lymphoid cells to divide, differentiate and produce specific antibody (serum and cell-bound). Antigen in non-phagocytosable form may lead to specific immune paralysis, perhaps by direct interaction with lymphoid cells.

Antibodies are produced only in vertebrates and in all species investigated have a fundamentally similar structure (Fig. 2) made up of two heavy and two light chains covalently linked by at least three interchain disulphide bonds (see Cohen and Milstein, 1967). The four-chain unit has a molecular weight of 150,000–180,000 and carries two combining sites, each of which comprises about 1 per cent of the total surface and is located in the N-terminal section of the molecule (Fig. 2). Human antibodies can be differentiated into three major classes which have common light chains and are distinguished by the chemical structure of their heavy chains; higher molecular weight antibodies, e.g. IgM, are polymers of the basic four-chain unit. Molecules in the three classes of antibody may have identical combining specificity, but their biological effects are modified by differences in distribution within the body fluids and by a varying capacity to fix complement and to be attached to cells and membranes (Table 1). Two additional classes of immunoglobulins have recently been described. IgD is present in low and variable concentration and has not been shown to carry antibody activity. IgE is also a trace component, but there is growing evidence that this class may, in some subjects, be associated with reaginic antibodies which mediate various immediate-type hypersensitivity responses in man (Stanworth, Humphrey, Bennich and Johansson, 1967).

Certain immunological responses cannot be passively transferred with serum, but are transferable with lymphoid cell suspensions. This is true for delayed hypersensitivity responses in general, e.g. the tuberculin reaction, and applies also to specific sensitivity to grafts of organized tissue (Mitchison and Dube, 1955). Such immunological phenomena, which appear to depend upon the direct interaction of antigen with sensitized cells, have led to the concept of 'cell-bound antibody' (Fig. 1). Whether such cells are passive carriers of absorbed antibody or actively synthesize the specific combining sites which are presumably carried at their surface, is at present unknown (see Coombs, 1967).

THE IMMUNE RESPONSE TO TISSUE ANTIGENS

It is now established that the rejection of genetically foreign tissue grafts is fundamentally dependent upon an immunological process evoked by transplantation antigens of the graft. If these antigenic molecules are actually released from the graft, as they appear to be from certain cultured cells (Mannick, Graziani and Egdahl, 1964), then the process of immunization would be similar to that outlined above for soluble or particulate antigens. However, the time of contact required for grafts to evoke sensitization and their failure to sensitize the host when contained in cell impermeable filter chambers suggest that tissue antigens may remain

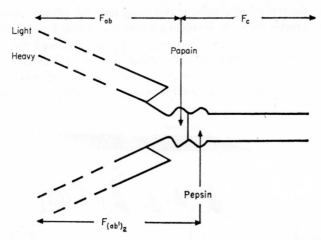

FIG. 2. Diagrammatic representation of IgC antibody. The molecule is made up of two heavy and two light chains linked by three inter-chain disulphide bonds. Broken lines show the highly variable portions of the molecule which contain the two antibody-combining sites. The undulating portion of the heavy chain represents the area susceptible to proteolytic digestion; the enzymes papain and pepsin give rise to the fragments indicated (from Cohen and Milstein, 1967).

TABLE I

Properties of human immunoglobulins

	IgC	IgA	IgM
Normal concentration (mg %)	800–1680	140–420	50–190
Molecular weight	150,000	170,000 400,000*	900,000
CHO content (%)	2.9	7.5	11.8
Heavy chains	γ	α	μ
Light chains	κ, λ	κ, λ	κ, λ
Biological properties			
Antibody activity	+	+	+
Complement fixation	+	o	+
Placental passage	+	o	o
Presence in CSF	+	+	o
Selective seromucous secretion	o	+ *	o

* The IgA in seromucous secretions appears to consist of two four-chain units associated with an addition 'Transport' (T) piece.

fixed to the transplanted cells. If this is so, then the afferent arc of the immune process probably involves sensitization of host lymphocytes by direct contact with foreign cells during their passage through the grafted tissue (Fig. 3). Such sensitized cells probably divide and differentiate in remote lymphoid tissues and liberate showers of specific effector lymphocytes (carrying cell-bound antibody) capable of invading and destroying the graft.

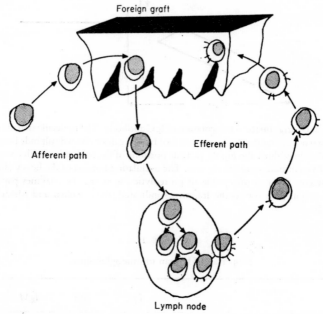

Foreign graft

Efferent path

Afferent path

Lymph node

Fig. 3. Diagrammatic representation of the afferent pathway for sensitization of host lymphocytes to foreign tissue, and the efferent pathway leading to graft rejection by 'cell-bound antibody'.

IMMUNOLOGICAL REACTIONS IN RELATION TO THE NERVOUS SYSTEM

Some sites in the body, notably the substantia propria of the cornea and the anterior chamber of the eye, have distinctive vascular connections which apparently render them relatively inaccessible to antibody and unlikely to liberate antigenic macromolecules into the general circulation. Certain experimental observations have suggested that the brain should be included among these immunologically 'privileged sites'; however, tissues of the central nervous system appear in some circumstances to be the source of antigenic material and, in addition, nervous tissue may become accessible to circulating antibodies.

(i) *Antibodies to nervous tissue—the afferent arc* (Fig. 4)
It has long been established that mammalian nervous tissue is antigenic and when injected into heterologous species evokes the formation of

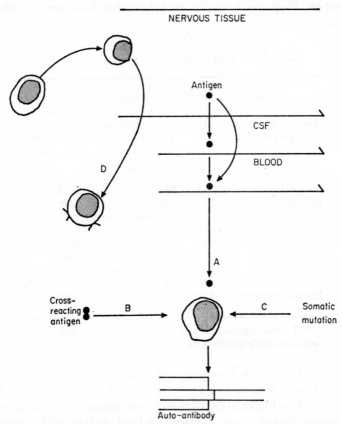

FIG. 4. Diagrammatic representation of various pathways which could lead to the production of antibody against components of the nervous system (see text).

specific complement-fixing antibodies. Comparatively little is known about the precise chemical nature and location of antigenic components within the nervous system, but various proteolipid, glycoprotein, protein and glycolipid fractions with antigenic properties have been studied (see Adams and Leibowitz, 1967). Detailed information is becoming available for the antigens involved in the pathogenesis of acute allergic encephalitis. These comprise a group of low molecular weight basic proteins localized in the myelin sheath (Kornguth and Anderson, 1965) and representing a small fraction of the total protein in white matter. Active protein fractions free of lipid have been isolated and shown to have molecular weights from 20,000 to 40,000 (Caspary and Field, 1963) to as low as 3000 (Robertson, Blight and Lumsden, 1962; Lumsden, Robertson and Blight,

TABLE 2

Antibodies to CNS components in various clinical states

Antigen	Test*	Positive in	Reference
Ganglioside	H	Multiple sclerosis	
Asialoganglioside	H	Tay Sachs' disease	Yokoyama
		Schizophrenia	et al., 1962
		Viral encephalitis	
Encephalogenic protein	H	Multiple sclerosis	Field et al.,
		GPI	1963
		Presenile dementia	
Trypsinized human brain	P	Multiple sclerosis	Rieder et al., 1964
Homologous or heterologous CSF	T	Multiple sclerosis	Fowler et al., 1966

* H = haemagglutination, P = precipitation, T = transformation of patient's lymphocytes.

1964, 1966; Carnegie and Lumsden, 1967). Three encephalogenic proteins with molecular weights of 16,000, 14,000 and 7000 were isolated by Nikao, Davis and Einstein (1966) from bovine spinal cord; these fractions are immunologically and chemically related and probably represent fragments of a single protein. An analogous protein fragment (molecular weight less than 8000) has been isolated from human white matter and myelin (Einstein, Csejtey and Davis, 1965).

If such antigenic material were liberated from nervous tissue it would presumably enter the circulation and gain access to cells of the immunological system (Fig. 4A). Thus, labelled proteins are able to penetrate the ependymal lining of the ventricles (Klatzo *et al.*, 1962, 1964) through intercellular spaces (Brightman, 1965) and protein present in the CSF readily enters the blood stream (Sweet, Selverstone, Solloway and Stetten, 1950; Lippincott, Korman, Lax and Corcoran, 1965). Since maturation of the immune system precedes the development of at least some nervous system antigens, it is to be expected that these would evoke an immune response on entering the circulation in the adult. This may account for the fact that antibodies to a variety of nervous system constituents (Table 2) have been observed in various neurological disorders and occasionally in normal subjects (reviewed by Adams and Leibowitz, 1967). Alternatively, since the classes of CNS compounds known to be antigenic are widely distributed in living organisms, these antibodies may have been formed against bacteria or viruses carrying cross-reacting antigens (Fig. 4B). Such antibodies could also be formed as a result of lymphoid cell mutation (Fig. 4C).

Finally, it seems likely that lymphocytes may under certain circumstances become sensitized by direct interaction with tissues within the central nervous system (Fig. 4D). The role of lymphocytes in the afferent arc of the tissue rejection process has been outlined above (Fig. 3). Earlier experiments showed that intracerebral transplants survived for unusually long periods; this suggested that the brain is an 'immunologically privileged' site, possibly because no interaction can occur with host lymphocytes (see Woodruff, 1960). In more recent experiments, however, the functional and morphological integrity of intracerebral grafts has been short-lived (Sheinberg, Edelman and Levy, 1964; Block, Tworek and Miller, 1966; Lance, 1968). In addition, animals with intracerebral thyroid grafts showed an accelerated rejection of subsequent skin grafts taken from the endocrine donor (Eichwald, Wetzel and Lustgraaf, 1966). This demonstrates that antigens within cerebral tissue can lead to generalized immunity, presumably as a result of direct lymphocyte sensitization.

(ii) *Antibodies to nervous tissue—the efferent arc*
Certain antibodies to nerve tissue have a direct cytolytic effect resulting in demyelination and damage to glial cells maintained in tissue culture. Such antibody, which is cytotoxic only in the presence of complement, has been found in human sera from cases of multiple sclerosis (Bornstein and Appel, 1961; Bornstein, 1963; Appel and Bornstein, 1964), amyotrophic lateral sclerosis (Bornstein and Appel, 1965) and motor neurone disease (Field and Hughes, 1965). Similar antibodies occur in sera from various species with experimentally induced allergic encephalitis (Bornstein and Appel, 1965; Lamoureux *et al*, 1966, 1967). The *in vivo* effect of

such potentially damaging antibodies is, of course, dependent upon their capacity to gain access to tissues of the central nervous system.

The production of cerebral damage by circulating antibody *in vivo* is strikingly illustrated by the injection into guinea-pigs of antibody against the Forssman antigen. This antigen is widely distributed in plants and animals. In the guinea-pig it occurs in many tissues including the vascular endothelium. Injection of this antibody into the vertebral artery leads to

TABLE 3

Total protein and Ig levels in serum and CSF (subjects with CSF total protein below 35 mg %) (Bannister and Cohen, unpublished)

	SERUM				CEREBROSPINAL FLUID			
	Total protein (g %)	IgG (mg %)	IgA (mg %)	IgM (mg %)	Total (protein) (mg %)	IgG (mg %)	IgA (mg %)	IgM
	7.0	1080	380	100	28	2.4	1.2	—
	6.3	1220	340	36	12	1.9	1.3	—
	4.9	1080	304	54	11	1.4	1.0	—
	6.2	1120	300	59	22	2.1	1.8	—
	7.6	740	138	160	8	1.2	1.3	—
	7.5	1100	144	34	19	2.7	1.2	—
	7.4	990	238	39	27	1.8	1.2	—
	6.9	760	120	80	22	1.3	1.3	—
	7.1	1220	520	48	15	1.6	1.3	—
	8.5	910	476	160	32	2.0	1.6	—
Mean	6.9	1022	296	77	20	1.8	1.3 9	—
	% Serum value				0.3%	0.2%	0.4%	0

destruction of the endothelium of cerebral vessels followed by widespread necrosis of nervous tissue (Leibowitz, 1966). Lesions due to a direct action of cytolytic antibody on extravascular cerebral tissues have not been demonstrated *in vivo*. This may be attributable in part to a restriction, either qualitative or quantitative, in regard to the passage of Ig molecules through cerebral capillaries. Thus, the concentration of IgG and IgA in CSF is less than 0.5 per cent of that in serum, while IgM is not detectable by immunochemical methods (Table 3). In this connection it is of interest that experimental allergic encephalitis cannot be passively transferred with serum unless this is injected directly into the ventricles (Jankovic, Draskoci and Janjic, 1965).

There is now considerable evidence to suggest that cell-bound antibody can act as the effector in immunological reactions directed against intra-cerebral tissues. For example, as discussed above, foreign grafts implanted in the brain undergo rejection by a process generally thought to be mediated by cells. In addition, experimental allergic encephalitis can be passively transferred in animals of an inbred strain by injection of lymph node cells but not readily with serum (reviewed by Adams and Leibowitz, 1967). The manner in which lymphocytes traverse capillaries to gain access to cerebral tissue remains obscure. The cellular infiltration character-istic of experimental allergic encephalitis appears to precede the occurrence of increased vascular permeability measured by accumulation of ^{131}I-albu-min in the brain (Leibowitz, 1966).

(iii) *Immunoglobulin production within the CNS*
Several observations indicate that immunoglobulin synthesis (and pre-sumably also antibody production) can occur within the central nervous system itself. In a variety of pathological states, including neurosyphilis, tuberculous meningitis and multiple sclerosis, there is a disproportionate increase of γ-globulin in the CSF (reviewed by Schultze and Heremans, 1966). In addition, IgM, which is not normally detectable, has been found in the CSF in many clinical disorders including neurosyphilis and benign lymphocytic meningitis. When CSF samples with raised levels of immuno-globulin are analysed by electrophoresis, the patterns and mobilities of γ-globulin are frequently quite distinct from those of serum. All these findings suggest that immunoglobulin may be produced locally within the central nervous system. This possibility is supported by the fact that in subjects with raised levels of CSF Ig, the specific activity of ^{131}I-labelled IgG injected intravenously remains considerably lower in CSF than in serum (Frick and Scheid-Seydel, 1958). More recently, the *in vitro* synthesis of both IgG and IgA, but not IgM, has been directly demonstrated using lymphocytes of the CSF from a patient with multiple sclerosis (Cohen and Bannister, 1967). It appears certain, therefore, that in patho-logical states immune responses may be confined to lymphocytic cells present in the CSF and the tissues of the nervous system.

(iv) *Inactivation of neurotrophic factors by specific antibody*
The fundamental discovery of Levi-Montalcini and her co-workers that growth and maintenance of sympathetic nerve cells is dependent upon a specific protein (nerve growth factor) has led to the demonstration that profound neurological lesions can result from inactivation of a neuro-trophic factor by specific antibody (reviewed by Levi-Montalcini, 1964;

C

Zaimis, 1965). Nerve growth factor which can be isolated in potent form from mouse salivary glands causes a remarkable hypertrophy of sympathetic ganglia when injected into newborn animals. Conversely, a specific antiserum to nerve growth factor administered to newborn animals results in widespread destruction of sympathetic ganglia (a process referred to as immunosympathectomy). The complexity of the nervous system and the sequential development of many of its structural components makes it likely that there are many other factors regulating its growth and differentiation. Whether the inactivation of any such factors by naturally occurring antibody can result in neurological lesions in human subjects remains to be determined.

COMMENT

The interaction of antigenic material with cells of the immunological system may lead either to the production of specific antibody (serum and cell-bound), or alternatively to a specific inhibition of the capacity to form that antibody (immunological paralysis). Paralysis is induced most readily in young animals and this reaction apparently provides the means whereby autologous tissues are recognized as 'self' and fail to evoke autoantibody formation. Antigens which develop after maturation of the immunological system may fail to induce paralysis and remain potentially antigenic in the adult. At least some antigenic component of the CNS, including those in the myelin sheath, probably fall into this category. This may account for the fact that antibodies to CNS components are found in a variety of pathological states and also in subjects without neurological lesions.

Most autoantibodies do not have the capacity to damage the tissues against which their combining specificity is directed, so that their role in the pathogenesis of disease is often in doubt. In the case of nervous tissue the fundamental work of Bornstein and his colleagues has shown that antibodies from patients with various neurological disorders are cytolytic when tested *in vitro* against myelinating cultures of nerve tissue. The capacity of such antibody to produce damage *in vivo* is of course dependent upon its ability to make direct contact with nervous tissue. The view that the CNS comprises an 'immunologically privileged' site does not appear to be tenable. Tissues of the nervous system can, it seems, be affected by immune processes originating in the general immunological system or within the brain itself; in addition, antibody to neurotrophic factors may lead to widespread damage of specific nervous tissues. The role of these various processes in human disease remains to be established, but their potential importance is indicated by experimental findings.

REFERENCES

ADAMS C.W.M. and LEIBOWITZ S. (1967) In *Structure and Function of Nervous Tissue* Vol. 3, Ed. BOURNE G.H. Academic Press, New York

APPEL S.H. and BORNSTEIN M.B. (1964) *J. exp. Med.* **119,** 303

BLOCK M.A., TWOREK E.J. and MILLER J.M. (1966) *Archs Surg. (Chicago)* **92,** 778

BORNSTEIN M.B. (1963) *Natn. Cancer Inst. Monogr.* No. 11, p. 197

BORNSTEIN M.B. and APPEL S.H. (1961) *J. Neuropath. exp. Neurol.* **20,** 141

BORNSTEIN M.B. and APPEL S.H. (1965) *Ann. N.Y. Acad. Sci.* **122,** 280

BRAUN W. and COHEN E.P. (1968) In *Regulation of the Antibody Response.* Ed. CINADER B., C. C. Thomas, Springfield, Illinois (in press)

BRIGHTMAN M. W. (1965) *J. Cell Biol.* **26,** 99

CARNEGI P.R. and LUMSDEN C.E. (1967) *Immunology* **12,** 133

CASPARY E.A. and FIELD E.J. (1963) *Nature (Lond.)* **197,** 1218

COHEN S. and BANNISTER R. (1967) *Lancet* **i,** 366

COHEN S. and MILSTEIN C. (1967) *Nature (Lond)* **214,** 449

COOMBS R.R.A. (1967) *Proc. R. Soc. Med.* **60,** 594

EICHWALD E.J., WETZEL B. and LUSTGRAAF E.C. (1966) *Transplantation* **4,** 260

EINSTEIN E.R., CSEJTEY J. and DAVIS W. (1965) *Proceedings of the 8th International Congress of Neurology, Vienna,* p. 137

FIELD E.J., CASPARY E.A. and BALL E.J. (1963) *Lancet* **ii,** 11

FIELD E.J. and HUGHES D. (1965) *Brit. med. J.* **ii,** 1399

FOWLER I., MORRIS C. and WHITLEY T. (1966) *New Engl. J. Med.* **275,** 1041

FREI P.C., BENACERRAF B. and THORBECKE G.J. (1965) *Proc. natn. Acad. Sci. U.S.A.* **53,** 20

FRICK E. and SCHEID-SEYDEL L. (1958) *Klin. Wschr.* **36,** 857

JANKOVIC B.D., DRASKOCI M. and JANJIC M. (1965) *Nature (Lond.)* **207,** 428

KLATZO I., MIGUEL J. and OTENASEK R. (1962) *Acta Neuropath. (Berl.)* **2,** 144

KLATZO I., MIGUEL J., FERRIS P.J., PROKOP J.D. and SMITH D.E. (1964) *J. Neuropath. exp. Neurol.* **23,** 18

KORNGUTH S.E. and ANDERSON J.W. (1965) *J. Cell Biol.* **26,** 157

LAMOUREUX G., BOULAY G. and BOUDAS A.G. (1966) *Clin. expl. Immunol.* **1,** 307

LAMOUREUX G., McPHERSON T.A. and CARNEGIE P.R. (1967) *Clin. expl. Immunol.* **2,** 253

LANCE E.M. (1968) *Transplantation* (in press)

LEIBOWITZ S. (1966) *J. Roy. Coll. Phycns (Lond.)* **1,** 85

LEVI-MONTALCINI R. (1966) *Harvey Lect.* **60,** 217

LIPPINCOTT S.W., KORMAN S., LAX L.C. and CORCORAN C. (1965) *J. nucl. Med.* **6,** 632

LUMSDEN C.E., ROBERTSON D.M. and BLIGHT R. (1964) *Z. ImmunForsch. exp. Ther.* **126,** 168

LUMSDEN C.E., ROBERTSON D.M. and BLIGHT R. (1966) *J. Neurochem.* **13,** 127

MANNICK J.A., GRAZIANI J.T. and EGDAHL R.H. (1964) *Transplantation* **3,** 321

MITCHISON N.A. (1968) In *Regulation of the Antibody Response.* Ed. CINADER B., C. C. Thomas, Springfield, Illinois (in press)

MITCHISON N.A. and DUBE O.L. (1955) *J. exp. Med.* **102,** 179

NIKAO A., DAVIS W. and EINSTEIN E.R. (1966) *Biochim. Biophys. Acta* **130,** 163

RIEDER H.P., ROSS J., RITZEL G. and WUTRICH R. (1964) *Med. exp. (Basel)* **11,** 128

ROBERTSON D.M., BLIGHT R. and LUMSDEN C.E. (1962) *Nature (Lond.)* **196,** 1005

SCHULTZE H.E. and HEREMANS J.F. (1966) In *Molecular Biology of Human Proteins*, p. 750. Elsevier, Amsterdam

SHEINBERG L.C., EDELMAN F.L. and LEVY W.A. (1964) *Archs Neurol. (Chicago)* **11,** 248

STANWORTH D.R., HUMPHREY J.H., BENNICH H. and JOHANSSON S.G.O. (1967) *Lancet* **ii,** 330

SWEET W.H., SELVERSTONE B., SOLLOWAY S. and STETTEN D. (1950) Studies of formation flow and absorption of cerebrospinal fluid. II. Studies with heavy water in the normal man, p. 376. American College of Surgeons Surgical Forum, Philadelphia. Saunders.

WOODRUFF M.F.A. (1960) In *The Transplantation of Tissues and Organs*, Chapter 3, p. 57. C. C. Thomas, Springfield, Illinois

YOKOYAMA M., TRAMS E.G. and BRADY K.O. (1962) *Proc. Soc. exp. Biol. Med.* **111,** 350

ZAIMIS E. (1965) *Proc. R. Soc. Med.* **58,** 1067

CHAPTER THREE

'AUTOIMMUNE' DEMYELINATING DISORDERS

K. J. Zilkha

Neurological sequelae of acute infectious fevers have been recognized for over 200 years, and it is 40 years since Bouman and Bok (1927) and Perdrau (1928) described perivascular demyelination in such cases. The first description of encephalomyelitis following smallpox was by Clifton (1724). It is now recognized that a similar disorder can follow measles, chicken pox, influenza, rubella and mumps, and may also follow Jennerian vaccination, the administration of anti-rabies vaccine and other prophylactic inoculations. The pathological changes, very similar in all these cases, has led to the use of such terminology as Acute Disseminated Encephalomyelitis, Post- or Para-infectious Encephalopathy and Acute Perivascular Myelinoclasis to describe this group of disorders. To some extent the histological picture may vary depending on the stage reached at the time of death. In the early stages there may be congestion and mononuclear cell infiltration of the walls of the small vessels, with subsequent oedema and haemorrhages (Fig. 1). After 3 or 4 days the picture changes to more intense perivenous infiltration with microglial and lymphocytic cells with subsequent demyelination in these areas. The picture is presumably reversible in the majority of cases, but in some cases demyelination may persist and lead on to more extensive gliosis (Fig. 2). Hurst (1941) separated off a clinical-pathological entity of acute haemorrhagic leucoencephalitis, usually following pharyngitis or upper respiratory tract infections, from other forms of haemorrhagic encephalopathy. Occasionally there may be no antecedent infection, and the onset is very abrupt with pyrexia and stupor or coma. There may be convulsions or hemiplegia and death results in a few days. Russell (1955) reported two cases following an influenza-like illness, which showed an intermediate pathological picture between acute haemorrhagic leucoencephalitis and post-infectious encephalitis. Glanzmann (1927) suggested that the disorders were in the nature of an immunological response. Despite repeated attempts to do so, claims to have obtained a virus from the brain or spinal cord in these cases have been exceedingly rare. Shaffer, Rake and Hodes (1942) obtained measles virus from the brain of a fatal case by inoculation into a monkey,

TABLE 1

The 'Autoimmune' demyelinating disorders

Terminology
Acute disseminated encephalomyelitis
Post- or para-infectious encephalopathy
Acute perivascular myelinoclasis
Acute haemorrhagic leuco-encephalitis

Causes
Post-vaccinial
    antirabies vaccine
    Jennerian vaccination
    antitetanus serum
    influenza vaccine
Post-infectious
    measles
    rubella
    chicken pox
    mumps
    scarlet fever
    whooping cough
    influenza

Review of multiple sclerosis

Diagnosis

Treatment

---

and Lopez Fernandez, Perez Sora and Ramirez Corria (1947) isolated the virus from the spinal fluid of a patient who recovered subsequently. In the great majority of cases no virus has been isolated. Moreover, as Glanzmann pointed out, these cases show a definite incubation time, a characteristic histological picture, a resemblance to experimental encephalomyelitis of animals and apparent personal idiosyncrasy.

POST-VACCINIAL

ANTIRABIES VACCINATION

Most authorities now accept that an immunological disorder is the most likely explanation for the group as a whole. This is readily acceptable in the case of the allergic encephalomyelitis or polyneuritis which may follow the administration of a course of injections of rabbit brain or spinal cord to individuals who have come in contact with or been bitten by a suspectedly rabid animal. Parish and Cannon (1962) give an incidence of 1 in 6000 persons given treatment developing a serious paralysis. Less serious neurological complications are of course more common. Sharp and McDonald (1967) reported their experience with the use of Semple vaccine, a killed virus suspension of rabbit brain origin. Twenty persons, airways and zoo personnel, who had close contact with a rabid leopard cub transported from Nepal to Edinburgh Zoo, were given the vaccine. Many side-effects were reported, with impaired concentration in 14, forgetfulness in 11, headaches of varying severity in 13 and 1 patient had severe unilateral arm weakness with sensory loss. This individual was given corticosteroid cover for the remainder of his course of vaccine, and was away from work for 3 months. Five other persons had transient paraesthesiae or weakness. This is, of course, a high incidence rate of side-effects—but active immunization against rabies remains current practice (Hildreth, 1963) and offers better protection than passive immunization with rabies antiserum, particularly to those whose exposure was some days before. Other vaccines, the duck embryo or the high-egg passage Flury vaccine, offer less protection in the post-infection treatment of guinea-pigs than the Semple vaccine (Veeraraghavan and Subrahmanyan, 1963). Moreover, neurological complications occur with other vaccines, too. Pathologically there may be quite large areas of congested white matter with greyish areas of demyelination round the small venules. In my own personal experience during army service overseas, I saw the complications of anti-rabies vaccination—including two patients with myelitis and one with encephalomyelitis—but no case of rabies. However, one patient died of rabies at the National Hospital nearly 3 years ago. She had arrived from Indonesia and the diagnosis was confirmed at autopsy. So long as the quarantine regulations for the admission of animals to this country remain strict, the problem will be a sporadic one. It may be debatable whether all persons receiving anti-rabies vaccine should also have ACTH or steroids but there is a stronger case for their administration in those who develop neurological complications. In parallel to the example of experimental allergic encephalomyelitis, prompt treatment with large doses is probably advisable.

VACCINATION AGAINST SMALLPOX

A big outbreak of post-vaccinial encephalomyelitis occurred in Holland in the years 1924–26, when 139 cases were reported, and there were forty-one deaths. Bouman and Bok (1927) reported the pathological findings on these cases, and Turnbull and McIntosh (1926) reported seven cases of encephalitis following vaccination against smallpox in this country. This is some 70 years after vaccination against smallpox was made compulsory in Britain. The incidence of this complication has shown wide variations in published reports from different countries and at different periods. As an example one may cite the experience in Glasgow during the 1942 outbreak of smallpox when 500,000 persons were vaccinated and the incidence of encephalitis was 1 in 70,000 (Anderson and McKenzie, 1942) whereas Dixon (1962) gives an incidence as high as one for every sixty-three people vaccinated in the small towns of Holland. These are, however, exceptional figures. In the outbreak of smallpox in South Wales in early 1962 there were forty-five cases with seventeen fatalities and some 800,000 were vaccinated. Spillane and Wells (1964) reported thirty-nine cases of a neurological complication attributed to the vaccination. Apart from the single fatal case, the authors personally examined all the other cases. Twenty-four had a primary vaccination and fifteen a revaccination.

There were thirteen cases of cerebral dysfunction—which Spillane and Wells (1964) divided into two groups, of encephalomyelitis and encephalopathy (Table 2A). The ten patients who presented acutely in the second week after successful vaccination with cerebral, brainstem and spinal cord symptoms or signs and were presumed to have the histological picture of microglial encephalitis as described by de Vries (1960), together with the one case of transverse myelitis, are grouped under encephalomyelitis. The three remaining cases of cerebral disturbance occurring within a month of vaccination are described as examples of post-vaccinial encephalopathy. They presented with epileptic seizures with subsequent development of focal neurological signs but made a full recovery in a matter of weeks.

The seven patients with meningism but no other signs had a raised cerebrospinal fluid pressure. The clinical picture was ascribed to the viraemia. Two patients had an isolated fit 9 and 6 days after vaccination. One patient had petit mal status, but she gave a history of an episode of confusion with incontinence of faeces 1 year earlier.

The next group of six patients with focal lesions of brain and spinal cord include two who were already suspected of having multiple

TABLE 2

Neurological complications of Jennerian vaccination (from Spillane and Wells, 1964)
Epidemic of smallpox in South Wales, January to May 1962.
Forty-five cases of smallpox, with seventeen deaths.
Number of persons vaccinated—over 800,000.
Thirty-nine had neurological complications.

Clinical picture	No. of cases	Comment
A. Central nervous system		
Encephalomyelitis	10	Two were in coma—one died. Clouding of consciousness in eight
Transverse myelitis	1	Rapid onset paraplegia with double incontinence. Recovery after 7 weeks
Encephalopathy	3	Epilepsy with hemiparesis or paraparesis. Two had status epilepticus and one had purpura and haematuria
Meningism of viraemia	7	Meningism only. No other signs
Epilepsy	3	Isolated fit with no sequelae in two, and petit mal status in one
Focal lesions of brain and spinal cord	6	Three had exacerbation of probable multiple sclerosis. Three with no previous neurological disturbance had a picture suggestive of multiple sclerosis
B. Peripheral nervous system		
Peripheral neuritis	5	All made a good recovery but one patient was left with facial palsy 6 months later
Brachial neuritis	2	
Myasthenia gravis	2	Relapse coincided with appearance of vaccination papule

sclerosis and another had had a previous neurological illness. The other three patients had multiple lesions and in one recovery did not take place until 2 months later.

The group of five patients with peripheral neuritis (Table 2B) includes one patient who made a full recovery but died suddenly of a massive pulmonary embolus 7 months later. There was no evidence of a previous encephalitis on examination of the brain. Both cases of brachial neuritis showed pain and weakness of an arm, the same as the vaccination side in one, and sensory loss in the territory of the circumflex nerve in this patient.

Both patients with myasthenia gravis who deteriorated a week after vaccination did so in dramatic fashion. One was suddenly so weak in the night that she could not reach the tablets by her side or waken her husband. The other developed difficulty in eating, swallowing and speaking overnight. Both improved to their previous clinical state, although the second patient was away from work for 6 months.

The cases of myasthenia and multiple sclerosis raise the question of desirability of vaccination in people with a known neurological disorder. McAlpine, Compston and Lumsden (1955) and Miller and Schapira (1959) reported single cases of multiple sclerosis who had an episode 14 and 2 days respectively following vaccination. My own practice is to advise patients with multiple sclerosis against vaccination and if necessary issue a certificate to that effect. Certified exemption on health grounds has been accepted by immigration authorities.

In cases of encephalomyelitis following Jennerian vaccination, the cerebrospinal fluid is usually abnormal at some stage of the illness, with both a lymphocytic pleocytosis and a moderate increase in the protein content. The EEG is also abnormal, with bilateral high voltage slow activity.

In a comprehensive monograph on post-vaccinial perivenous encephalitis de Vries (1960) describes foci of perivenous demyelination with microglial proliferation throughout the white matter in both hemispheres. There is no coalescence and no softening at the centre of the foci. He considers the condition as a specific disease of the vaccinial non-immune. But similar cases have been reported following successful revaccination and de Vries cites three such cases. Apart from his twenty-nine cases with such a picture, de Vries describes another group of thirty-three cases as post-vaccinial encephalopathy with a greater incidence of convulsions, and more severe lesions with increased likelihood of residual disabilities in those who recover from the acute illness. It is interesting to note that twenty-four of the thirty-three cases were aged 4 months to 2 years, whereas the youngest with encephalitis was 3 years old. The three cases of encephalopathy in the series of Spillane and Wells (1964) were aged 6, 8 and 11 years.

Consideration of the neurological complications of vaccination raises the controversial question of the practice of Jennerian vaccination at the present time. This is not the occasion to go into the pros and cons of a policy of universal vaccination. Suffice it to say that there are many factors to be considered, including the likelihood of foreign travel to countries who insist on the production of a valid certificate of vaccination against smallpox, and the nature of the individual's work, for instance in an acute fever hospital which may admit a case of smallpox at any time. Moreover, it has been suggested (Wynne Griffith, 1959; Dixon, 1962; Dick, 1962) that primary vaccination in adult life carries a smaller mortality than primary vaccination in infancy with revaccination during or after adolescence. Further evidence is obviously needed, but it does question the present practice of advising vaccination against smallpox in the second year of life.

TREATMENT AND PROPHYLAXIS

ACTH and steroids
There are reports of the use of steroid hormones in cases of encephalo-myelitis following vaccination against smallpox (Ligterinck, 1951; Nossel and Rabkin, 1956). In the series of Spillane and Wells (1964) six patients were treated with ACTH, one was given prednisolone and one patient who had been in a coma for 10 hours before admission to hospital received intravenous hydrocortisone but died 24 hours later. This was the only fatality. Another patient who was also in coma and had choreiform movements and double incontinence was given large doses of ACTH, 240 units/day. He was fully orientated 3 days later, and was fully recovered and at home after 2 weeks.

Spillane and Wells (1964) make the point that in three of their cases recovery set in within 12 hours from commencing ACTH therapy.

Gamma-globulin
In a controlled trial of anti-vaccinia gamma-globulin (AGG) among military recruits in Holland, Nanning (1962) reports an incidence of three cases of post-vaccinial encephalitis out of 53,630 receiving 2 ml of 16 per cent AGG at the time of primary vaccination, compared with an incidence of thirteen cases of post-vaccinial encephalitis out of 53,044 receiving a placebo at the time of primary vaccination. This difference is significant. The figures quoted give an incidence of 1 in 4000 primary vaccinations

for the incidence of this disorder in adults, similar to the one given for pre-school children in Holland. Revaccination involves a much lesser risk, in the region of 1 in 50,000.

ANTITETANUS SERUM

Cases of isolated neuritis or radiculitis following the injection of anti-tetanus serum (Allen, 1931) are more common than acute disseminated encephalomyelitis which has been reported only infrequently (Miller and Stanton, 1954). Miller and Ramsden (1962) report the case of a girl of 23 who received 1 ml of antitetanus serum following an accident resulting in a fracture of the femur and abrasions. On the fifteenth day she had acute retention of urine and, subsequently, flaccid paraplegia, weakness of the left arm, unequal pupils and sensory loss below the second dorsal segment. She deteriorated rapidly and became drowsy and confused. She died 9 days later, the day after intravenous steroids were first administered. At autopsy the brain showed diffuse haemorrhages in the white matter and the spinal cord showed acute necrosis of the white matter with intense congestion of the leptomeninges. Histologically there was extensive demyelination, with haemorrhages, cellular infiltration, fibrinoid necrosis of blood vessel walls and extravasation of fibrin—features of acute dis-seminated encephalomyelitis and of haemorrhagic leucoencephalitis.

INFLUENZA VACCINE

Reports of neurological sequelae following the administration of influenza vaccine are infrequent (Warren, 1956; Woods and Ellison, 1964). The case reported by Woods and Ellison (1964) was that of a 7-year-old girl who was given a subcutaneous injection of 0.1 ml of polyvalent influenza vaccine consisting of types A and B influenza virus grown in the extra-embryonic fluids of the developing chick embryo and inactivated by formaldehyde. Within 24 hours she developed fever, lethargy and vomit-ing. Over the next few days she complained of frontal headache and vomited intermittently but suddenly lapsed into coma on the eighth day. She showed blurring of one optic disc, sluggish reflexes with extensor plantar responses. She recovered consciousness after a few hours, was confused and disorientated for the next 3 days but recovered fully by the seventeenth day. She did not receive steroid therapy. Apparently the child had had influenza vaccine 2 or 3 years before. Neurological complications have also been described following TAB, anti-diphtheria and triple vaccine (pertussis, diphtheria and tetanus) inoculation (Miller and Stanton, 1954).

POST-INFECTIOUS

Reports of the incidence of neurological complications in the specific acute infections give widely divergent figures. Experience in different

TABLE 3

Incidence, latent period and mortality of encephalomyelitis and polyradiculitis associated with the specific fevers (after review by Miller, Stanton and Gibbons, 1956)

	Incidence	Average latent period		Clinical picture (% of all cases)		Mortality (%)
Measles						
911 cases	1:1,000	4.7 days	E	96	E	27
		5.5 days	M	3	M	20
		?	P	1	P	0
Rubella						
80 cases	1:5,000	3.8 days	E	93	E	20.3
		6.7 days	M	3.5	M	33.3
		8.7 days	P	3.5	P	33.3
Chicken pox						
121 cases	?1:1,000	6.3 days	E	90	E	10
		8.3 days	M	3	M	0
		11.3 days	P	7	P	0
Mumps						
40 cases	?	7.2 days	E	68	E	22.2
		10.5 days	M	17	M	0*
		11.3 days	P†	15	P	33.3
Scarlet fever						
30 cases	1:1,500–	8.7 days	E	73	E	13.6
	1:10,000	0	M	7	M	0
		0	P	20	P	0
Whooping cough						
97 cases	? less than 1:100	(2–4 weeks)	E	100	E	

E = Encephalitis or encephalomyelitis.
M = Myelitis.
P = Polyradiculitis or polyneuritis.

* Other reports include some fatalities.
† Some present before the appearance of parotitis.

countries, and in different epidemics in the same country, shows that the incidence is variable. Of course, this information does depend on the

completeness of national notification both of the neurological complication and of the specific fever.

In an excellent and critical review, Miller, Stanton and Gibbons (1956)
analyse the results of a survey of the literature of the neurological complications of the individual specific fevers up to 1953. As these authors rightly
point out, no clinician sees more than a very limited number of such
cases, and I have used their comprehensive review as the basis of my
remarks.

MEASLES

The incidence of encephalitis complicating measles is probably in the
region of one case in a thousand. The average latent period is just under 5
days, but rarely may antedate the rash by as much as 10 days or come on 3
weeks after. Mild and severe cases of measles are equally liable to neurological sequelae, and gamma-globulin does not seem to confer any protection from the neurological complications. The encephalitic illness may
begin abruptly with a fit or there may be gradual lapse into coma. There is
usually headache, vomiting and, frequently, disturbed behaviour, restlessness and irritability. Drowsiness and high fever are almost always
present. The reported mortality is in the region of 20–30 per cent and
survivors usually make a good recovery in a matter of a few weeks.
However, rarely, there may be residual neurological deficit.

Case report

B.F. (K.C.H. C 84101), aged 9 years, was admitted to hospital with a
history of measles complicated by encephalitis at the age of 3 years. As a
result he lost his speech and walking ability and had a spastic tetraparesis.
His I.Q. at the age of 8 years was estimated to be 73 on W.I.S.C. Ten
days before his final admission to hospital he developed an upper respiratory tract infection which was treated with intramuscular penicillin
with clinical improvement. The day before admission he complained of
generalized limb pain, vomited and became oliguric. He showed severe
spastic tetraparesis. The next day he developed oculogyric crisis and
myoclonic jerkings. Cerebrospinal fluid examination was normal. He had
haematuria and later developed circulatory failure, had a large haemate-
mesis and died 3 days after admission and 3 weeks after the onset of the
final illness. At autopsy there was severe haemorrhagic congestion of the
lungs, acute fatty change in the liver and congestion of the kidneys.

The brain showed extensive damage to the central white matter which
was represented by very few myelinated axons (Fig. 3). There was no
damage to either cortex or other grey matter.

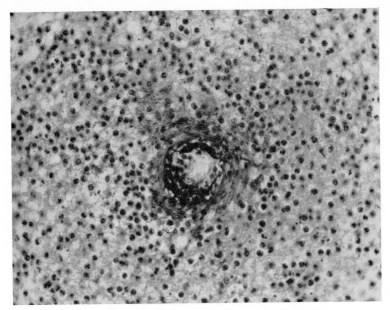

FIG. 1. Marked perivascular mononuclear cell infiltration.

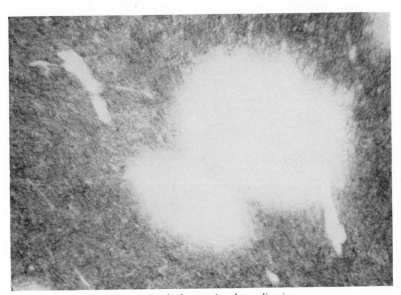

FIG. 2. Marked extensive demyelination.

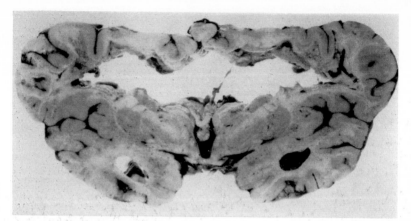

FIG. 3. Brain of a boy aged 9 years who had measles, complicated by encephalitis at the age of 3, with residual tetraparesis. There is extensive damage to the central white matter which is represented by very few myelinated axons.

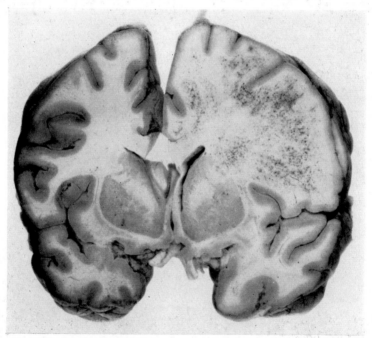

FIG. 4. Brain of a man aged 25 years who died from acute haemorrhagic leucoencephalitis. There are numerous petechial haemorrhages in the central white matter of the left hemisphere and the left side of the corpus callosum.

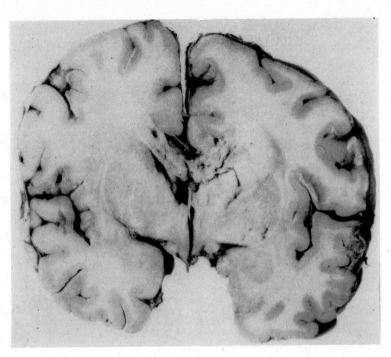

FIG. 5. Brain of a man aged 32 years who died with acute necrotizing leuco-encephalitis. There is widespread acute inflammatory and necrotizing processes involving the white matter with heavy infiltration with polymorphs, marked oedema, perivascular exudation of fibrin and severe perivascular demyelination.

FIG. 6. Biopsy from man of 50 years with leucoencephalitis showing area of rarefaction in the white matter, centred round a blood vessel.

The improvement in the neurological picture may still leave behind a personality disorder, and Ramsay and Young (1964) describe a permanent change in three children following severe encephalitis with prolonged loss of consciousness and fits. Relapse or recurrence of measles encephalomyelitis must be very rare, if it occurs at all, whereas recurrence is not infrequent in acute disseminated encephalomyelitis which may follow upper respiratory tract infections (Alcock and Hoffman, 1962).

In a much smaller number of patients the clinical picture is that of a myelitis (3.0 per cent) and polyradiculitis (1.0 per cent). In myelitis the onset may be abrupt, with backache, retention of urine and weakness and sensory loss associated with a transverse or ascending cord lesion, and uncommonly there may also be bulbar signs. The lesion in myelitis may progress for a few days before gradual recovery sets in. The comparatively high mortality of 20 per cent in the review by Miller, Stanton and Gibbons is associated with bulbar involvement, in the course of an ascending myelitis. The smaller group with polyradiculitis also present with weakness of the legs, with accompanying painful paraesthesiae. In nine out of the ten patients in this review there were also bulbar symptoms, with dysphagia, dysarthria and diplopia. There were no deaths, and in the majority recovery was gradual over many months.

RUBELLA
Although in a particular epidemic neurological complications may appear in more than 1 per cent of the cases, generally speaking these are less frequent than in measles, with an incidence of about 1 case in 5000. The latent period is usually about 4 days, but the neurological complication may preceed the rash by a few days or come on up to 2 weeks after. In the review by Miller, Stanton and Gibbons, the majority of cases, 93 per cent, were examples of rubella encephalitis, and the remainder examples of myelitis or polyradiculitis (Table 3). Convulsions and loss of consciousness are the commonest presenting signs while other cases may present with headache, vomiting and delirium. In about a third of the patients, the onset may be more gradual with increasing drowsiness or meningeal signs. The mortality in the cases reviewed was in the region of 20 per cent, and about 38 per cent of comatose patients died, after a brief illness of less than 3 days. The prognosis in survivors is good. The more rare cases of myelitis may present with an ascending paralysis with transverse lesion of the cord. The equally rare cases of polyradiculitis present with a flaccid tetraparesis. Death may occur from bulbar involvement. Tinel and Bénard (1923) reported the finding of perivascular

demyelination running the whole length of the affected vessels in the spinal cord.

CHICKEN POX

The neurological complications resemble those of measles and rubella. The incidence is not known and is probably similar to that in measles, 1 per 1000. The latent period varies and may be up to 3 weeks after the appearance of the rash although, again rarely, neurological signs may appear before the rash. The severity of the neurological illness does not appear to be related to the severity of the rash or the length of the incubation period. Here, again, encephalitis is the most frequent neurological complication in the reported cases (90 per cent).

The onset is again variable, and convulsions and sudden coma may herald the illness in a dramatic way. Almost as frequently, patients may complain of headache, and meningism may appear. Varying degrees of drowsiness may be present, and coma in about 10-20 per cent of cases. The mortality is in the region of 10 per cent, and survivors generally have a good prognosis. However, residual deficits which have been reported include cerebellar ataxia, retarded development and convulsions, ophthalmoplegia and optic atrophy. The few cases of ascending myelitis reported all presented between the sixth and tenth day after the appearance of the rash. All these cases made a complete recovery after 3-6 weeks. Patients showing the picture of polyradiculitis developed a flaccid weakness of the limbs, with bulbar signs in about half of them. There were no deaths in the cases reported, and recovery was usually complete 4 months after the onset.

MUMPS

The commonest neurological complication of mumps is, of course, meningitis. Cases of mumps myelitis and meningoencephalitis may be the result of the virus invasion of the central nervous system, and perhaps spreading from the meninges to the cerebrum and spinal cord. However, some cases do appear to resemble the clinical picture of acute disseminated encephalomyelitis complicating the other specific fevers. The latent period ranges from 5 days before to 3 weeks after the parotid gland becomes enlarged, with an average of about 1 week. In their review Miller, Stanton and Gibbons included twenty-seven such cases, presenting acutely in twenty, with meningism in nine, convulsions in eight and one went into coma. The other seven patients included four who lapsed into stupor slowly, and other three had weakness of the legs, increased muscle tone and hemiparesis. Six of the twenty-seven patients died, a mortality of 22.2 per cent. Of the survivors, seven patients (one-third) showed major sequelae including hemiplegia, severe ataxia and tremor in the short follow-up period. Donohue (1941) reviewed the pathological reports of mumps encephalitis and noted that the undoubted cases had shown peri-venous encephalitis. Cases of mumps myelitis, as in the other fevers, may present with a transverse cord lesion; or ascending paralysis, paraesthesiae and urinary retention may be the presenting symptoms. The cases of poly-radiculitis present with paraesthesiae and flaccid weakness of the limbs, and usually cranial nerve involvement, particularly the facial nerve. Deafness may also occur. Death from respiratory failure has been reported. The prognosis in survivors is good, with recovery in the majority of cases.

SCARLET FEVER

The neurological complications of scarlet fever are very diverse. Some may be associated primarily with otitis or nephritis which may follow streptococcal infection. Meningitis and hemiplegia are not infrequently reported. However, cases of encephalitis, myelitis and polyradiculitis also occur and the incidence is in the region of 1 in 5000. The cases of scarlatina encephalitis may present acutely with convulsions; coma and vomiting may be prominent. Increasing drowsiness may be the salient sign of these cases. Hemiplegia is an uncommon presentation, and in the review by Miller, Stanton and Gibbons there was no such example. The latent period was again variable from 2 days before to 3 weeks after the appearance of the rash.

The mortality is higher in the cases with coma or generalized convulsions. Sequelae are more common following scarlet fever encephalitis and one-fifth of the cases may be left with cerebellar ataxia, mental retardation and spasticity. Cases of myelitis are rare. Miller and Evans (1953) reported the case of a man of 37 years with motor and sensory signs below the waist which came on whilst the rash was fading. In the cases of polyradiculitis the latent period tends to be longer, and up to 6 weeks after the appearance of the rash. All cases show hypotonia and loss of tendon reflexes. Prognosis is usually good.

WHOOPING COUGH OR PERTUSSIS

The neurological complications of pertussis differ from those discussed above in that the clinical picture follows a distinct pattern. Generalized convulsions, coming on 2 weeks or so after the onset of paroxysms, are

D

followed by coma and frequently by focal neurological signs. The commonest focal disturbance is hemiplegia, and less commonly aphasia and monoplegia. Pathological reports are rare and although Möller (1949) described perivenous encephalitis, there is still uncertainty as to the exact nature of these cases, and anoxia or haemorrhagic change may be responsible in part. Miller, Stanton and Gibbons therefore prefer the term encephalopathy to describe the complications following pertussis.

UPPER RESPIRATORY TRACT INFECTION AND INFLUENZA

Cases of encephalitis and encephalomyelitis following upper respiratory tract infection are probably just as common as those already considered. Headache, drowsiness and lethargy may follow a few days after the onset of a feverish illness characterized by catarrh and general aches and pains which had already begun to show signs of improvement. Convulsions and stupor with more definite muscular weakness herald the complication. More acute cases may present in coma, either immediately or after a few days of the onset of the influenzal illness. In one such case examined by Greenfield (1958) influenza A virus was obtained from the lungs but not from the brain. The pathology was that of acute haemorrhagic leuco-encephalitis.

Case reports

Case one. W.C. (N.H. 84516), a man aged 25 years, was admitted with a history of recurrent headaches for 7 years. He was diagnosed as suffering from migraine and X-rays of the skull and an EEG, done elsewhere, were normal. A week before his admission he developed a cold with a cough, productive of yellow sputum. The day prior to admission he vomited repeatedly and was incontinent of urine. On admission he appeared to be alert and answered questions by movements of the head. He had no speech. He was right-handed. There was no obvious weakness of the limbs but the tendon reflexes were more brisk on the left side with bilateral extensor plantar responses. He collapsed and died a few hours later. Examination of the brain showed numerous petechial haemorrhages in the central white matter throughout the frontal and parietal lobes of the left hemisphere, the left half of the corpus callosum, and in the internal capsule on the left side (Fig. 4).

Case two. S.B. (M.H. 1372), a man aged 32 years, was admitted to hospital with a week's history of a cold and cough which had improved. The night before his admission he vomited repeatedly and lapsed into coma. He had a temperature of 102° F, was sweating profusely with some neck rigidity and small petechial haemorrhages on both legs. There were no localizing signs on examination of the central nervous system. CSF examination showed a turbid fluid with 5330 polymorphs and 330 lymphocytes per cmm. The protein content was 335 mg per 100 ml. He was treated with intrathecal penicillin, I.V. sulphadimidine followed by I.M. chloramphenicol. He remained unconscious and died 36 hours later. Examination of the brain showed greying of the white matter more marked over the left hemisphere but with striking sparing of U. fibres. The corpus callosum was extensively involved. Microscopically there was a widespread acute inflammatory and necrotizing process involving the white matter with very heavy infiltration with polymorphs, marked oedema, marked perivascular exudation of fibrin and severe demyelination, perivascular in distribution. The cortex was relatively unaffected (Fig. 5).

Recurrent attacks of encephalomyelitis have been reported by van Bogaert, Borremans and Couvreur (1932), van Bogaert (1950) and Alcock and Hoffman (1962). It is notable that the predisposing factors may be different in the subsequent attacks from the initial episode. In van Bogaert's two cases the first (van Bogaert et al., 1932), a boy of 7, showed a pyramidal lesion after anti-scarlet fever serum at the age of 5, with a recrudescence after a further attack of scarlet fever at 6½ years and who finally succumbed 4 months later to another neurological illness complicating measles, with cerebellar and pyramidal signs and coma. The second (van Bogaert, 1950) was a girl of 9 years who developed polyneuritis with diphtheria at the age of 6, diplopia, ataxia, choreiform movements and bilateral extensor plantar responses following an attack of measles 2 months later. Three years later she had a feverish illness followed by diplopia, optic neuritis, flaccid tetraplegia, facial weakness and bilateral extensor plantar responses, and died after 14 days. The brain showed widespread perivascular demyelination, and cellular infiltration of the white matter.

Alcock and Hoffman (1962) described six cases who survived recurrent episodes of encephalomyelitis in childhood. They advocate corticosteroid therapy for the acute epidosdes.

REVIEW OF MULTIPLE SCLEROSIS

From a consideration of recurrent episodes of encephalomyelitis in children, I should like next to review briefly some of the problems of multiple sclerosis.

Is multiple sclerosis an autoimmune disease, resulting from immunization against the patient's own nervous tissue? There are many who hold this view. Of course, the typical pathological lesion, the plaque, is quite unlike anything seen in the cases of acute disseminated encephalomyelitis which follow infections or inoculations. However, Uchimura and Shiraki (1957) have reported the finding of large demyelinated plaques, similar to those of multiple sclerosis, in the brains of a number of patients who died some months after the advent of neurological complications or anti-rabies vaccinations. Berg and Kallen (1962), using a tissue culture technique reported the finding of a 'gliotoxic' factor in the serum in twenty-seven of forty-two cases of multiple sclerosis, but also in other neurological conditions mainly with traumatic and vascular brain lesions. Bornstein (1963), using a myelin culture technique, reported the finding of a 'myelino-toxic' factor in the serum of fourteen of twenty cases with 'active' exacerbation of multiple sclerosis but not in the fifteen 'inactive' cases. However, eight of fifty cases of other neurological disorders also gave positive results. The nature of the raised gamma-globulin in CSF is still uncertain. Is it an autoimmune antibody, or is it an incidental product of tissue breakdown? The fractional pattern of the gamma-globulin in the CSF in multiple sclerosis seems to correspond with that of the chronic infections such as syphilis, or allergic disorders such as post-vaccinial encephalitis and polyarteritis nodosa. This may suggest an immunological nature (Lumsden, 1965). Another widely held view is that multiple sclerosis is due to a virus with a long latent period (Sigurdsson, Palsson and Grinsson, 1957; Poskanzer, Schapira and Miller, 1963) even though attempts to obtain a virus from the nervous system have failed. Raised titres of measles antibody have been reported (Adams and Imagawa, 1962; Reed, Sever, Kurtzke and Kurland, 1964; Clarke, Dane and Dick, 1965) but so far there has been no supporting evidence from immuno-fluorescence studies. Others have put foward biochemical and vascular hypotheses; and Thompson (1966) reviewed the reported changes in serum lipids with a decrease in linoleate content and increased platelet stickiness in the acute exacerbations in this disease. More recently Caspary, Sewell and Field (1967) reported increased red blood cell fragility in multiple sclerosis.

This is not the place to review all the pros and cons of these not necessarily conflicting views. Any solution to the problem of the aetiology of multiple sclerosis will have to explain the many known clinical factors in this disease, including its age incidence, familial occurrence, the temporal relation of exacerbations to infections, trauma, and the geographical incidence (McAlpine, Lumsden and Acheson, 1965; Symonds, 1959). There are cases on record when the disease appeared to have its onset closely following inoculation with TAB/TT (Miller and Schapira, 1959). The cases described by Spillane and Wells (1964) following Jennerian vaccination have already been mentioned. One patient with multiple sclerosis seen at the National Hospital developed an acute paraplegia a few days after the administration of 'Russian' vaccine, which was really very similar to anti-rabies vaccine and prepared from the formalinized suspension of rat brain infected with 'Sv' virus (Shubladze and Gaidamovitch, 1956). Examples of patients who have had acute exacerbations following banal infections are not uncommon in clinical practice.

While the aetiology of multiple sclerosis remains in doubt, there have been reports of the beneficial effects of corticotrophin in multiple sclerosis, in the short term and in the acute relapses (Miller, Newell and Ridley, 1961; Alexander and Cass, 1963; Rawson, Liversedge and Goldfarb, 1966).

On the other hand, Millar, Vas, Nogonha, Liversedge and Rawson (1967) found no evidence of any such improvement in the long term in a large number of patients when compared with controls studied over the same period of time.

DIFFERENTIAL DIAGNOSIS

It is not infrequent that patients with acute cerebral symptoms of disseminated encephalomyelitis or haemorrhagic leucoencephalopathy should be considered to have a space-occupying lesion such as an abscess, haematoma or tumour and that verification of the diagnosis should be sought from a cerebral biopsy.

Case report
P.E. (N.S.U. 4371), a man aged 50 years, was admitted to hospital on 21.3.64. The previous day whilst at work he had general malaise, and in the evening developed weakness of the left leg and he became drowsy. He was apyrexial. In hospital a history of minor head injury a few weeks previously was elicited and the diagnosis of a subdural haematoma was suspected. ESR was 30 mm (Westergren)/hour, WBC 18,000 per cmm. A right carotid arteriogram showed slight depression of the middle cerebral artery and a shift of the anterior cerebral artery to the left of the midline. The picture was of an expanding right frontal lesion, extending deeply and posteriorly. There was no evidence of a subdural haematoma. A right frontal craniotomy was performed on 25.3.64—the gyri were flattened

and brain tissue felt softer than normal on palpation. Biopsies were taken. The histological picture was that of numerous areas of rarefaction in the white matter, centred round blood vessels (Fig. 6). In these areas there were fat granule cells and polymorphs. He started improving after 2 days. Two months later he had made a complete recovery. It is interesting that subsequently it was learnt that the patient was given one prophylactic injection against influenza on 28.11.63 about 4 months prior to his illness. Whether this is relevant or coincidental, it is difficult to be sure.

In general, however, a history of infection, signs of bilateral cerebral lesions or more widespread disturbance of function should lead to the suspicion of the diagnosis. The EEG may show bilateral abnormalities. Spinal fluid examination is unlikely to be of help in the differential diagnosis from cerebral abscess.

Patients presenting with a predominantly spinal lesion may be suspected of having an acute exacerbation of multiple sclerosis. However, long-term follow-up of these cases helps to differentiate the two groups.

GENERAL REMARKS ON TREATMENT

Corticotrophin and corticosteroids have been used in all these disorders. Successful suppression of experimental allergic encephalomyelitis (EAE) in various animals by the timely administration of ACTH or corticosteroids has been repeatedly reported (Moyer, Jervis, Black, Koprowski and Cox, 1950; Kabat, Wolf and Bezer, 1952). Partial suppression, using smaller doses of ACTH or corticosteroids has also been reported (Field and Miller, 1962; Fois, Pieri and Malandrini, 1959). However, in all these reports the administration of ACTH or corticosteroids was begun at the time of the inoculation or a week before the inoculation. Gammon and Dilworth (1953) reported good results using larger doses of ACTH at the time of the onset of the disease. Using larger doses of methyl prednisolone administered the day after some of the animals showed signs of EAE and 10 days after inoculation, Kibler (1965) reported complete clinical suppression of EAE in the treated rabbits, so long as the high dosage was maintained, and deterioration when the dose was drastically reduced. While, therefore, there seems to be a strong case for using large doses of either corticotrophin or corticosteroids in these cases of auto-immune demyelinating conditions, there is not sufficient evidence to recommend one rather than the other. In the case of corticotrophin it has been shown that in the dog, and probably in man, maximal secretion rates for cortisol and corticosterone are reached with doses that do not produce maximal aldosterone secretion (Ganong, 1963) (Fig. 7). This gives some justification for using doses in excess of 100 units of corticotrophin daily for an adult.

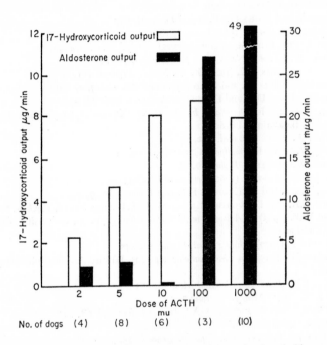

FIG. 7. Effect of ACTH on 17-hydroxycorticoid (open bars) and aldosterone (blocked bars) output in nephrectomized, hypophysectomized dogs. Values are adrenal venous outputs.

ACKNOWLEDGMENTS

I wish to thank Dr Mary Wilmers, Dr Michael Kremer and Mr J. J. Maccabe for permission to publish details of patients under their care. I also wish to thank Dr S. Strich, and Dr W. Mair for providing the histological pictures.

I am grateful to the editors and publishers for permission to reproduce tables and figures from Brain, *The Quarterly Journal of Medicine*, and *Neuroendocrinology*.

REFERENCES

ADAMS J.M. and IMAGAWA D.T. (1962) *Proc. Soc. exp. Biol. Med.* **111,** 562

ALCOCK N.S. and HOFFMAN H.L. (1962) *Archs Dis. Childh.* **37,** 40

ALEXANDER L. and CASS L.J. (1963) *Ann. internal Med.* **58,** 454

ALLEN I.M. (1931) *Lancet* **ii,** 1128

ANDERSON T. and McKENZIE P. (1942) *Lancet* **ii,** 667

BERG O. and KÄLLÉN B. (1962) *Lancet* **i,** 1051

BOGAERT L. VAN, BORREMANS and COUVREUR (1932) *Presse méd.* **40,** 141

BORNSTEIN M.B. (1963) *Nat. Cancer Inst. Monogr.* **11,** 197

BOUMAN L. and BOK S.K. (1927) *Z. ges. Neurol. Psychiat.* **111,** 495

CASPARY E.A., SEWELL F. and FIELD E.J. (1967) *Brit. med. J.* **2,** 610

CLARKE J.K., DANE D.S. and DICK G.W.A. (1965) *Brain* **88,** 953

CLIFTON F. (1724) In *Dissertatio Inauguralis de Distinctis et Confluentibus Variolis.* Lugduni Batavosum, Leyden

DICK G.W.A. (1962) *Brit. med. J.* **2,** 1275

DIXON C.W. (1962) In *Smallpox.* Churchill, London

DONOHUE W.L. (1941) *J. Pediat.* **19,** 42

FIELD E.J. and MILLER H. (1962) *Brit. med. J.* **1,** 843

FOIS, A., PIERI I. and MALANDRINI F. (1959) *Pediatria Int.* **9,** 503

GAMMON G.A. and DILWORTH M.J. (1953) *A.M.A. Archs Neurol. Psychiat.* **69,** 649

GANONG W.F. (1963) In *Advances in Neuroendocrinology,* p. 96. University of Illinois Press, Urbana

GLANZMANN E. (1927) *Schweiz. med. Wschr.* **57,** 145

GREENFIELD J.G. (1958) In *Neuropathology,* p. 205. Edward Arnold (Publishers) Ltd. London

HILDRETH E.H. (1963) *Ann. internal Med.* **58,** 883

HURST E.W. (1941) *Med. J. Australia* **2,** 1

KABAT E.A., WOLF A. and BEZER A.E. (1952) *J. Immunol.* **68,** 265

KIBLER R.F. (1965) *Ann. N.Y. Acad. Sci.* **122,** 469

LIGTERINCK J.A.T. (1951) *Ned. Tijdschr. Geneesk.* **95,** 3490

LOPEZ FERNANDEZ F., PEREZ SORA E. and RAMIREZ CORRIA F. (1946) *Revta Méd. Cirug. Habana* **52,** 385

LUMSDEN C.E. (1965) In *Multiple Sclerosis, a Reappraisal,* Chapter XI, p. 298. E. & S. Livingstone Ltd, Edinburgh and London

McALPINE D., COMPSTON N.D. and LUMSDEN C.E. (1955) In *Multiple Sclerosis.* E. & S. Livingstone Ltd, Edinburgh and London

McALPINE D., LUMSDEN C.E. and ACHESON E.D. (1965) In *Multiple Sclerosis, a Reappraisal.* E. & S. Livingstone Ltd, Edinburgh and London

MILLAR J.H.D., VAS C.J., NORONHA M.J., LIVERSEDGE L.A. and RAWSON M.D. (1967) *Lancet* **ii,** 429

MILLER A.A. and RAMSDEN F. (1962) *J. clin. Path.* **15,** 314

MILLER H.G. and EVANS M.J. (1953) *Q. Jl Med.* **22,** 347

MILLER H., NEWELL D.J. and RIDLEY A. (1961) *Lancet* **ii,** 1120

MILLER H. and SCHAPIRA K. (1959) *Brit. med. J.* **1,** 737

MILLER H.G. and STANTON J.B. (1954) *Q. Jl Med.* **23,** 1

MILLER H.G., STANTON J.B. and GIBBONS J.L. (1956) *Q. Jl Med.* **25,** 427

MÖLLER F. (1949) *Acta med. scand., Suppl.* **232**

MOYER A.W., JERVIS G.A., BLACK J., KOPROWSKI H. and COX H.R. (1950) *Proc. Soc. exp. Biol. Med.* **75**, 387

NANNING W. (1962) *Bull. Wld Hlth Org.* **27**, 317

NOSSEL H.L. and RABKIN R. (1956) *S. Afr. med. J.* **30**, 492

PARISH H.J. and CANNON D.A. (1962) In *Antisera, Toxoids, Vaccines and Tuberculins in Prophylaxis and Treatment*, 6th ed., p. 221. E. & S. Livingstone Ltd, Edinburgh and London

PERDRAU J.R. (1928) *J. Path. Bact.* **31**, 17

POSKANZER D.C., SCHAPIRA K. and MILLER H. (1963) *J. Neurol. Neurosurg. Psychiat.* **26**, 368

RAMSAY A.M. and YOUNG S.E.J. (1964) *Public Health* **78**, 100

RAWSON M.D., LIVERSEDGE L.A. and GOLDFARB G. (1966) *Lancet* **ii**, 1044

REED D., SEVER J., KURTZKE J. and KURLAND L. (1964) *Archs Neurol. (Chicago)* **10**, 402

RUSSELL D.S. (1955) *Brain* **78**, 369

SHAFFER N.F., RAKE G. and HODES A.L. (1942) *Am. J. Dis. Child.* **64**, 815

SHARP J.C.M. and McDONALD S. (1967) *Brit. med. J.* **3**, 20

SHUGLADZE A. and GAIDAMOVITCH S. (1956) In *Congresso Internazione di Patologia Infettiva*, Fasc. 2, 173

SIGURDSSON B., PALSSON P.A. and GRINSSON H. (1957) *J. Neuropath. exp. Neurol.* **16**, 389

SPILLANE J.D. and WELLS C.E.C. (1964) *Brain* **87**, 1

SYMONDS C.P. (1959) In *Biochemical Aspects of Neurological Disorders*, p. 46. Blackwell, Oxford.

THOMPSON R.H.S. (1966) *Proc. R. Soc. Med.* **59**, 269

TINEL J. and BÉNARD R. (1923) *Revue neurol.* **39**, 310

TURNBULL H.M. and McINTOSH J. (1926) *Br. J. exp. Path.* **7**, 181

UCHIMARA I. and SHIRAKI H. (1957) *J. Neuropath. exp. Neurol.* **16**, 139

VEERARAGHAVAN N. and SUBRAHMANYAN T.P. (1963) *Bull. Wld Hlth Org.* **29**, 323

VRIES E. DE (1960) In *Postvaccinial Perivenous Encephalitis*. Van Nostrand, Amsterdam

WARREN W.R. (1956) *A.M.A. Archs intern Med.* **97**, 803

WOODS C.A. and ELLISON G.W. (1964) *J. Pediat.* **65**, 745

WYNNE GRIFFITH G. (1959) *Brit. med. J.* **1**, 1343

DEMYELINATING DISEASES AND ALLERGIC ENCEPHALOMYELITIS. A COMPARATIVE REVIEW WITH SPECIAL REFERENCE TO MULTIPLE SCLEROSIS

R. E. Caspary

In recent years the causation of many diseases, especially those not completely understood by their investigators, has been placed into the category of 'autoimmune'. It may at this stage be of value to consider the exact meaning of this term and then to justify its application to nervous disease on the basis of proven experimental facts. Perhaps the only concrete example of true antibody-mediated autoimmune disease in man is acquired haemolytic anaemia, where an antibody to red cells, sometimes only those carrying specific known red cell antigens, can be shown to cause drastic reduction of their survival *in vivo*. The antibody, a γ-globulin, can be eluted from the patient's red cells and its specificity is then easily demonstrated. Antibody bound to tissues may be demonstrated in a number of disease states, although in this situation it would be more correct to speak of bound γ-globulin, unless specificity was also shown. In diseases such as disseminated lupus erythematosus and rheumatoid arthritis the antinuclear factors and anti-globulins have no proven pathogenetic effects and serum from patients with these diseases does react with a variety of tissues and tissue extracts in *in vitro* complement-fixation tests (Mackay and Gajdusek, 1958). Serum from normal individuals also frequently has 'autoantibodies' to tissues, usually without pathology; low titre antibody to thyroglobulin can be found in a very large proportion of the population and in the search for anti-brain antibody in disseminated sclerosis very low titre antibody to some components of this organ was found in some normal control individuals. Circulating antibody to the target organ, unless shown to be capable of transferring the disease process, does not then provide evidence of certain autoimmune disease. The initiation of such antibody production may be an injury to the target organ, releasing breakdown products ultimately into the reticuloendothelial system, when the normal protection against this event may be the structural integrity of the organ (Mackay and

Burnet, 1963) or in some cases a breakdown in immune tolerance could occur. Experimental allergic disease, lesions in a target organ following the injection of either whole tissues or purified tissue antigens together with adjuvants, is only slightly species specific; thus allergic encephalomyelitis may be induced in the guinea-pig with nervous tissue from man, ox or guinea-pig with similar efficiency. This does suggest that perhaps injury of any type may be the most likely initiating step in the activation of the immune mechanism and that this mechanism is functioning normally. The role of circulating antibodies in these experimental allergic conditions appears to bear little direct relation to the onset or severity of disease and passive transfer of serum has never produced lesions in the host. On the contrary, in allergic encephalomyelitis passive transfer of serum was shown to protect against or delay the onset of disease (Harwin and Patterson, 1962). Circulating antibody even to a specific disease producing antigen in these diseases can only provide an indication of an active immune process and direct attention to the delayed-type of sensitivity and the associated specifically sensitized cells. Delayed sensitivity may be transferred to another host by passive transfer of cells and experimental allergic disease has been passed in this way (Patterson, 1960). It is also possible that both antibody and sensitized cells may be required to obtain allergic disease perhaps in association with complement, but the mechanism of autoimmune tissue injury is as yet not understood. The ultimate demonstration of a true autoimmune disease then rests on the demonstration of circulating antibody capable of transferring disease by passive transfer or on the presence of demonstrable delayed hypersensitivity to the putative disease antigen(s); the action of antiserum or sensitized cells or both in the presence and absence of complement on tissue cultures of the target organ may be of the greatest importance in the assessment of human disease of possible 'autoimmune' pathogenesis.

Autoimmunity, or perhaps better autoallergy, as a unifying mechanism of a group of demyelinating diseases comprising multiple sclerosis, acute disseminated encephalomyelitis and acute necrotizing haemorrhagic encephalopathy has proved an attractive hypothesis to both experimental pathologists (Alvord, 1966) and clinicians (Miller and Schapira, 1959). It must be stated that other mechanisms, including vascular thrombosis (Ribbert, 1882; Putnam, 1935), vascular obstruction due to aggregated fat particles (Swank, 1950), spirochaetes (Kuhn and Steiner, 1917, 1920; Steiner, 1952) and virus infection, have been suggested at various times. However, there appears to be little evidence in favour of most of these hypotheses and choice is at present restricted to autoimmunity and virus infection. To date, no virus specific to central nervous tissue has been seen in the demyelinating diseases in spite of many careful searches, nor indeed is it accurate to refer to the disease produced by JHM virus (Cheever, Daniels, Pappenheimer and Bailey, 1949) or Visna (Sigurdsson and Palsson, 1958) as primary demyelinating conditions (Field, 1966). In addition, material from plaques of multiple sclerosis has no deleterious effect on nervous tissue in culture (Lumsden, 1955) and *in vivo* transfer of multiple sclerosis to Icelandic sheep (Palsson, Pattison and Field, 1965) and mice (Field, 1966) still awaits confirmation. Much attention has been centred on presumptive 'slow' virus infections of the central nervous system, such as 'Kuru' in the Fore tribe in New Guinea and 'Scrapie' and 'Visna' infections in sheep, and progress in these studies has been reviewed in a recent monograph (Gajdusek, Gibbs and Alpers, 1965). On the other hand, viral action does not exclude an allergic element in the disease mechanism, it would indeed provide an initiating mechanism for an auto-allergic reaction. At present auto-allergy still presents the most attractive hypothesis of the causation of demyelinating diseases following on the suggestion of Glanzmann (1927) in post-exanthematous encephalitis and extended later to multiple sclerosis for reasons of pathological affinity mainly by van Bogaert (1932) and Pette (1942). It may thus be said that the morbid anatomy of these diseases favours the allergic theory (Ferraro, 1944) and this point is extended by Adams and Kubik (1952) who also emphasize some common features in the whole group of demyelinating diseases and suggest that some property of the myelin may determine the localization of the pathology. 'Allergic' neurological diseases following the injection of nervous tissue has been known since before the turn of the century and in 1933 Rivers and others described an encephalomyelitis accompanied by myelin destruction in monkeys following the injection of rabbit brain extracts over a period of more than 6 months. With the advent of Freund's adjuvant (Freund and McDermott, 1942) it became possible to induce disease with a single injection of brain and adjuvant and allergic encephalomyelitis could be induced in a number of different animal species. Most mammalian central nervous tissue is capable of inducing this 'allergic' disease in both homologous and heterologous host-tissue relation and is thus organ- but not species-specific. As allergic encephalomyelitis may be produced with portions of the animal's own brain the experimental disease, and, by virtue of morbid anatomical similarity, the demyelinating disorders in man have been included in the now popular family of autoimmune diseases. However, only in the neuroparalytic accidents in man following the Pasteur treatment of rabies

(Uchimura and Shiraki, 1957), a true parallel to experimental 'allergic' encephalomyelitis in laboratory animals, do we find a picture resembling the common lesions of multiple sclerosis. This may be related to the species of animal used and the general rate of the pathological action but does suggest that experimental allergic encephalomyelitis may not be a strictly accurate model for multiple sclerosis in man. Histopathological study of the experimental disease showed it to be a necrotizing encephalitis (Pette, Mannweiler and Palacios, 1961) and electron microscopy showed that though myelin is destroyed this is often preceded by degenerative changes in the axis cylinders and glial cells are also damaged (Field and Raine, 1966). Therefore, to speak of allergic encephalomyelitis as a demyelinating disease places a particular emphasis on only one of its features (Field, 1966) and even the studies in man after the injection of nervous tissue (Uchimura and Shiraki, 1957; Jellinger and Seitelberger, 1958) though strongly supporting an immunological pathogenesis for multiple sclerosis do not give absolute histopathological identity with human disease (Lumsden, 1961). At this stage a detailed examination of the pathological mechanism at work in experimental allergic encephalomyelitis may provide further information on the similarities and differences between the experimental model and human demyelinating disease.

Allergic encephalomyelitis (EAE) shows substantial variations both in clinical and histological response in different species. The clinical response in the mouse is uncertain until paralysis occurs and there is evidence that genetic constitution has a profound influence on the disease in this species (Böhme, Lee, Schneider and Wachstein, 1966). In the guinea-pig neurological disturbance consisting of weakness or paralysis of the hind limbs, tremor, incontinence of urine usually accompanied by a sudden and severe weight loss occurs about 2 weeks after the challenge with foreign tissue; in rhesus monkey the signs appear around 1 week later consisting of irritability, signs of cerebellar ataxia, nystagmus, weakness in an arm or leg and apparent blindness. The neurological signs in the monkey may follow one another or fluctuate in a transient manner and, as in other experimental species, terminate fatally. In the smaller mammals histology in EAE shows patchy lymphocytic meningitis, infiltration of the tela choroida and disseminated perivenous accumulation of lymphocytes sometimes with plasma cells; rhesus monkey also shows widely disseminated lesions of perivenous demyelination, most commonly sited in the white matter but frequently extending into the grey. EAE in the monkey also shows polymorphonuclear leucocyte infiltration of the early stages of these necrotic lesions followed by the appearance of compound granular corpuscles; these areas of acute necrosis heal with relatively little glial scarring (Field, 1966). In those animals which survive an attack of EAE and whose tissues are not examined until a prolonged period of time has elapsed, few if any histological signs are seen and further challenge with active encephalitogen fails to produce a second attack of the acute disease. In general, most experimental work has been done on the rat or guinea-pig, permitting the statistical assessment of the results of various treatments using a combination 'score' based on both clinical signs and the histological appearance at a fixed time after the initial challenge (Kies, Goldstein, Murphy, Roboz and Alvord, 1957).

Much attention has been paid to the antigen responsible for the induction of EAE, though the disease may be adequately produced with homogenized suspensions of whole cord or brain or isolated white matter mixed with adjuvants containing mycobacteria and oils. It may readily be shown that the whole tissue contains a large battery of antigens of different specificities mostly shared with a number of other organs and of a variety of chemical constitution. Some may be specific to brain but not essential to the production of EAE and circulating antibodies produced will have many individual reactions not associated with the allergic disease. It has long been known that the white matter (myelin) is more active than the grey and that similarly immature brain (before myelination) injected as the whole tissue will not produce EAE, and work has been directed to the clarification of this point. The fractionation of brain tissue into subcellular particles by homogenization and ultracentrifugation on sucrose gradients has given definite evidence that encephalitogenic activity is concentrated in the myelin (Laatsch, Kies, Gordon and Alvord, 1962) and it could be suggested that the remaining activity may be due to contamination of other fractions with myelin fragments or even the encephalitogen itself released during the experimental manipulation.

The chemical approach to the isolation of encephalitogen is based essentially on the pioneer work of Folch, Lee, Kies, Roboz-Einstein and Alvord, with the emphasis moving from lipids to proteolipid and now to small proteins and peptides. Encephalitogenic activity is stable to extreme changes in pH. For instance, there is no loss in 0.02 N hydrochloric acid, it is not affected by solvents such as ether or even methanol-chloroform, which will dissolve myelin, and is also completely resistant to heat, showing no change in activity even after being autoclaved at 30 lb for 6 hours. However, digestion with trypsin and prolonged storage under very alkaline conditions does reduce encephalitogenicity of purified encephalitogen. On the other hand, under special conditions the factor will withstand the action of papain, producing a dialysable active fragment (Kies, Thompson and Alvord, 1965), whilst in whole tissue the activity is very resistant to proteolytic enzymes (Caspary and Field, 1963).

There is now general agreement that the encephalitogenic factor is protein in nature and belongs to the group of basic proteins related to, but not necessarily identical with, histones. These fractions have been isolated by acid extraction from defatted guinea-pig brain by Kies and Alvord (1959) and Kies, Shaw, Fahlberg and Alvord (1960) and also from bovine spinal cord by salt extraction by Roboz-Einstein, Robertson, Di Caprio and Moore (1962), and similar fractions were also isolated from human brain (Caspary and Field, 1965). In brief, encephalitogen is prepared by extraction of freeze-dried brain with methanol-chloroform to remove the lipids, washing out the solvents with water and removing further protein contaminants with 5 per cent sodium chloride. The residue is then extracted with 0.01 N hydrochloric acid at a pH of 2.5 or lower and the protein in the extract precipitated with ammonium sulphate at half saturation. The precipitate is dissolved in water, dialysed to remove salt and freeze dried to give a white water-soluble powder. The final product may be further purified by column chromatography on 'Sephadex' on an apparent size basis or by other chromatographic techniques based on ion exchange. It is of interest that chromatography on 'Sephadex' without special treatment to neutralize the basic charge effect gives totally erroneous estimates of molecular size. Fractions of apparent molecular weight of over 100,000 were found to have sedimentation coefficients of less than 1, corresponding to an approximate molecular weight of nearer 20,000 (Caspary and Field, 1965). The question of size of these basic proteins having encephalitogenic activity remains open. It appears that different methods of isolation may produce slightly different products and that even on standing in solution some degeneration of the basic protein takes place, usually without loss of biological activity. Robertson, Blight and Lumsden (1962) have isolated a dialysable peptide of very high biological activity from basic protein by alkaline dialysis and suggest that this peptide is the essential encephalitogen. A similar product is also described by Nakao and Roboz-Einstein (1965), although the question of molecular size and dialysability still remains open (Kies, 1965). Additional studies on the basic polypeptide (Lumsden, Robertson and Blight, 1966) state that all encephalitogenic activity is related to a small basic polypeptide of a molecular weight of about 5000 and that this may be bound to either basic proteins or a number of other constituents of the central nervous system of different chemical nature. The question remains as to the origin of the basic protein, carrying without doubt the highest biological activity, and the only direct evidence on this point remains the study of Laatsch and others (1962) on its isolation from purified myelin in high yield. It would be an attractive idea if the basic protein was shown to be the binding system between the sphingolipid complexes of myelin, thus providing some explanation of the 'demyelination' occurring in EAE in the higher mammals. One further interesting propery of encephalitogenic basic protein is its ability to react with nucleic acid, giving a firm complex of no biological activity, suggesting that perhaps the biological activity may be related to affinity for cellular nucleic acids (DNA or RNA) and that prior combination with foreign nucleic acids may mask the specific groupings (Kies *et al*, 1965). In addition, other work has shown that the basic protein has an equal or even greater affinity for DNA than histones from the central nervous system (Caspary, unpublished). Further work is in progress to relate these studies to the action of encephalitogen on the nucleic acids of the reticulo-endothelial system and the production of both circulating antibody and disease.

Encephalitogenic basic protein may be defined at this stage as a material of molecular weight of about 20,000, containing fifteen different amino acids (nitrogen 16–17 per cent) and some glucose, galactose and ribose (sugar 3 per cent), as well as a small amount of hexosamine, and perhaps having its specific activity in a basic peptide of similar chemical composition but of only 5000 molecular weight.

One further variable in EAE arises as a consequence of the era of adjuvant (Kabat, Wolf and Bezer, 1947; Morgan, 1947), as only in combination with these mixtures can the disease be produced rapidly and regularly with only one injection. However, this early advantage is not without its drawbacks. The components of the adjuvant (oil, emulsifier and mycobacteria) may vary from one laboratory to another and their relative proportion and composition has been shown to be critical in relation to experimental disease (Lee and Schneider, 1962; Shaw, Alvord, Fahlberg and Kies, 1962). In addition, the intense granulomatous reaction at the site of injection and the draining lymph nodes tends to obscure the tissue reaction to the nervous tissue under investigation. The inhibitory effect of adjuvant containing mycobacteria on EAE will be discussed at another stage. Early workers did produce EAE after multiple injection of brain or cord material but with a very poor experimental take (reviewed by Levine and Wenk, 1965), and only in very susceptible strains of rat could EAE be produced regularly and reproducibly without the aid of adjuvant.

The other method frequently used instead of adjuvants is priming the animals before challenge with encephalitogen with an injection of pertussis vaccine. This, too, introduces some strain on the immune mechanism but not the violent effect of complete Freund's adjuvant at the injection site and draining nodes. Other information arising from this work again places emphasis on the genetic make-up of the experimental animals and perhaps on the debit side that the quantity of specific central nervous tissue required to induce EAE even in very susceptible strains is many times greater than that needed in conjunction with adjuvant (Levine and Wenk, 1965).

Examination of the cerebrospinal fluid for cells, proteins and indeed any other constituents plays a considerable role in the diagnosis of neurological disease in man but is only possible in the larger experimental animals when comparing results between human and 'model' diseases. In EAE only the early work of Kabat and his colleagues showed increased γ-globulin in the cerebrospinal fluid of monkeys, but most later experiments have been done on the smaller mammalian species where these measurements are not technically feasible. In man with multiple sclerosis the spinal fluid γ-globulin is raised to a pathological level in 50–60 per cent of cases, making no distinction between the numerous sub-groups of this class of proteins. The total protein, too, shows a slight elevation in a smaller percentage of patients (Rieder and Wüthrich, 1962). The examination of the γ-globulin fraction in the cerebrospinal fluid of multiple sclerosis patients has been the subject of a large number of studies using widely differing techniques summarized in detail by Lumsden (1965). There is good agreement in the relatively crude paper electrophoretic methods, with only minor quantitative differences in the final mean globulin levels. This method and the even more crude routine chemical precipitation technique (Papadopoulos, Hess, O'Doherty and McLane, 1959) are those used as routine procedures in most clinical laboratories. More detailed studies using the much more sophisticated immunoelectrophoresis methods capable of detecting as many as thirty-six components in cerebrospinal fluid showed that only minor changes occurred in components other than the 'immunoglobulins' in neurological disease, especially in multiple sclerosis (Dencker and Swahn, 1961). These results, then, confirm that the increase is in the broad group of γ-globulin and does not arise from the sudden elevation of some hitherto minor or even new component. The protein described by MacPherson (1962) as specific to the cerebrospinal fluid does not appear to be adequately increased to account for the pathological elevation of globulin in disease. It remains to consider the origin and consequences of

E

this general increase of globulin with immune potential in the cerebro-spinal fluid, whether it is antibody and if so, to which antigen. Immune reactions in gel-diffusion systems, using either antisera against specific serum protein (Rosenthal and Soothill, 1962) or purified γ-globulin fraction from serum and cerebrospinal fluid (Caspary, 1965), indicated the immune identity of γ-globulins derived both from normal and multiple sclerosis cerebrospinal fluid and from serum. Frick and Scheid-Seydel (1958), in their studies on the derivation of globulins in the central nervous system using radioiodine-labelled protein, suggest that the increase in total globulin arises within the subarachnoid space on the basis of distribution of labelled and unlabelled protein in serum and cerebro-spinal fluid and this observation is in keeping with the well-known reluctance of antibody to penetrate into the undamaged brain. On the other hand, the antibody ratio in the serum and spinal fluid following the injection of some vaccines is altered in multiple sclerosis, suggesting some impairment in the blood brain barrier. This, however, is most likely to have occurred as a consequence of existing nervous disease and may only influence the disease process as a continuation mechanism. A reduction in one of the components of complement correlating with increased γ-globulin does occur in multiple sclerosis (Kuwert, Firnhaber and Pette, 1964), which points towards the presence of an immune process, but does not directly explain the increase in globulin. Ridley (1962) was able to show quantities of γ-globulin in the perivascular mononuclear cuffs in EAE in the guinea-pig, but similar studies have so far not been possible in human pathological material. A further possibility exists in the fact that non-antibody γ-globulin is produced following antigenic stimulus (Askonas and Humphrey, 1958) and inability to demonstrate antibody may be a consequence of all or most of it being bound in the tissue. It is possible that this may also account for the failure of Field and Ridley (1960) with the antiglobulin neutralization test, since the quantity of antibody sought may have been too small for the sensitivity of the test system. Thus, the spinal fluid γ-globulin most probably arises in the subarachnoid space in multiple sclerosis but its antibody nature and specificity remains in doubt at the present time.

Occurrence of complement-fixing antibodies to brain extracts in EAE in the rat was shown to correlate within a given group with the presence of lesions (Paterson, 1960), but this correlation did not exist in individual animals. Similarly, complement-binding antibody to purified encephali-togen was shown to be present in animals challenged with encephalitogen and adjuvant (Caspary, Sinden and Field, 1966), again not individually correlated with disease. This rise is accompanied by an increase in S19 globulin and this antibody is that described by Harwin and Patterson (1962) as being protective in rats. In man there have been numerous studies of complement-binding antibodies to nervous tissues, using a wide variety of preparations of tissue and an equally diverse number of detection systems. Of these, seven groups of investigators were able to show antibody in more than 20 per cent of cases of multiple sclerosis, whilst a further eight failed to find antibody at all or only in a small number of cases. All these studies were carried out with nervous-tissue suspensions, not encephalitogens, and it is possible that the positive reactions obtained may have been to components in the antigen unrelated to allergic disease or not even specific to brain. Indeed, there is evidence that the saline brain extracts used in the studies of Roberts (1962) and Caspary, Field, McLeod and Smith (1963) are not encephalitogenic (Caspary and Field, 1963) and yet gave diametrically opposite results by very similar complement-fixation techniques in multiple sclerosis. The divergence of results in this antibody study involving complement can at present only be interpreted as vaguely suggestive evidence in favour of some autoimmune disturbance.

An attempt to search for increases in immunoconglutinin, a variant of complement (Coombs, Coombs and Ingram, 1961), which is raised in known 'autoimmune' diseases such as Hashimoto's thyroiditis and rheumatoid arthritis, failed to give significant results either in EAE in the guinea-pig or in acute and chronic multiple sclerosis. There is, however, a changed distribution of immunoconglutinin titres in EAE rising towards higher titres, but the increase was not statistically significant. Similarly, in multiple sclerosis the titres and distribution were entirely normal in contrast to the 'autoimmune' diseases tested at the same time (Ball and Caspary, 1963; Caspary and Ball, 1962).

Delayed-type skin sensitivity to homologous and heterologous brain antigens in EAE has been reported in rabbits (Waksman, 1956). Only the heterologous material have been noted in the guinea-pig. However, Shaw, Alvord, Kaku and Kies (1965), using a purified encephalitogen, were able to produce true delayed-type skin reactions in the guinea-pig, being maximal between 9 and 13 days after challenge. Further work with encephalitogenic proteins and peptides gave essentially the same results (Lamoureux, Carnegie, McPherson and Johnston, 1967). The size of the reaction is claimed to correlate with the time of onset of the disease and the authors point out some of the differences between animals sensitized with homologous and heterologous preparations. The skin reaction diminishes when the animals become ill and test dosage is fairly critical. In a series of experiments using human antigen in guinea-pig no true delayed-type sensitivity was demonstrated against the encephalitogen, but positive results were obtained with an extract having no biological activity (Caspary and Field, 1965). This difference may have arisen on the technical grounds already mentioned. There have also been some attempts to study skin sensitivity in patients with multiple sclerosis (Stauffer and Waksman, 1954; Bauer and Heitman, 1958; Böhme, Paal, Kersten and Kersten, 1963) again with variable results and impure antigens, i.e. normal brain suspensions or crude fractions. Tests with encephalitogen failed to show any positive reaction in multiple sclerosis as did an inactive brain extract (Caspary and Field, 1965). These findings cannot be taken as excluding an immune pathogenesis as skin sensitivity and sensitization of the nervous system need not necessarily run parallel and there is evidence that this relation need not hold true (Swithinbank, Smith and Vollum, 1953; O'Grady, 1956).

Passive haemagglutinin reaction with brain extracts has been examined by two different techniques—that of chemical coupling with bis-diazobenzidine and the more common tanned cell method. Animals with EAE have antibodies by this method, but these again are only consistent within groups and do not apply to individuals. In multiple sclerosis titres are also significantly raised to brain extracts, but not to the same extent as that found in the serum of patients with rheumatoid arthritis or Hashimoto's thyroiditis (Caspary, Field and Ball, 1964). The experiments of Yokoyama, Trams and Brady (1962) also showed the presence of antibody to brain ganglioside, but Ahrengot (1957) found none against human white matter. When sera were tested with encephalitogen in the tanned cell system positive results were obtained both in EAE in the guinea-pig and in chronic and acute multiple sclerosis (Field, Caspary and Ball, 1963), but antibody to this antigen was also present in presenile dementia, general paralysis of the insane and a miscellaneous group of neurological diseases and also in a group of individuals recently vaccinated with BCG. It is of interest to note that in distinction to experiments with whole nervous tissue the 'model' autoimmune diseases, Hashimoto's thyroiditis and rheumatoid arthritis failed to give a significant response. These antibody reactions are not affected by treatment of serum with dimercaptopropanol, to split the complement-reacting S19 globulin, and therefore come into the category of S7 (IgG) globulin.

Ross (1960) and Ritzel, Wüthrich and Rieder (1963) have obtained significantly positive results by the use of trypsin digests of normal white matter in the agar diffusion test in a number of patients with multiple sclerosis. However, the specificity of the antigen reaction in this precipitin system is of some doubt, and some doubts have been expressed on the value of these results (Honegger, Ritzel and Rieder, 1964). Several authors have also claimed positive precipitin reactions in EAE, which in animals sensitized with an antigen adjuvant mixture is more likely to produce this type of antibody. More consistent positive precipitin results have been obtained by Lamoureux, Carnegie, MacPherson and Johnston (1967) using both encephalitogenic protein and peptides in EAE. Only the peptide of bovine origin failed to react, while the same fraction derived from guinea-pig nervous tissue was positive.

The extremely sensitive variation of the precipitin technique using radio-labelled antigen (Farr, 1958) has been applied in EAE by Kibler and Barnes (1962) in the rabbit and to EAE in the guinea-pig and multiple sclerosis by Caspary (1966), giving positive results in experimental disease but no response in multiple sclerosis. These authors used purified encephalitogens and results again indicate a group but not individual correlation with disease.

Passive transfer of the experimental disease with serum from animals with EAE has never been demonstrated. On the contrary, Harwin and Patterson (1962) were able to protect against encephalitogenic challenge by transfusion of serum from sick animals. The efficacy of this protection was related to the complement-binding titre of the serum against brain suspensions. Transfer with the use of lymphocytes or lymph node suspension appears to have been more successful in isologous strains of animals; transfer in guinea-pigs (Stone, 1962) and rats (Patterson, 1960) has been reported. A more recent study (Levine, Wenk and Hoenig, 1967) also recorded successful transfer of EAE in various genetic rat strains and was able to relate the effectiveness of transfer to the transplantation antigens in the system. Other reports of cellular transfer of EAE have made use of various techniques to suppress cellular rejection including splenectomy, X-irradiation and cytotoxic drugs and it is possible that some of these methods may ultimately lead to an experimental system capable of transferring human 'autoimmune' nervous disease of animals.

Drug suppression of EAE has been shown in rats and guinea-pigs with varying degrees of effectiveness—6-mercaptopurine has given variable results, as the borderline between toxicity and maximum effect in animals is regrettably close (Hoyer, Condie and Good, 1960; Field and Miller, 1961; Thomson and Austin, 1962), and the beneficial effect of corticosteroids both in EAE and multiple sclerosis has been reviewed by Kibler (1965). More recently cyclophosphamide in doses carrying little or no toxicity was shown to inhibit EAE in rats, achieving this inhibition even when treatment was started as late as 9 days after challenge with a potent encephalitogenic mixture (Patterson, Hanson and Gerner, 1967). The production of complement-binding antibody is also inhibited by cyclophosphamide. This drug is equally effective in the guinea-pig (Field, unpublished), but a clinical trial of cyclophosphamide in patients with multiple sclerosis is not as yet showing any inhibitory effect on the course of the disease.

Perhaps the most significant findings in the experimental disease and in multiple sclerosis is the ability of serum from cases of these disorders to cause demyelination of nervous tissues in culture reviewed by Lumsden (1965), Patterson (1966), Field (1966) and Hughes and Field (1967). Most of the experimental work in this system is based on the basic conditions determined by Hild (1957) and Bornstein and Murray (1958) and, in general, differences between different groups of investigators have derived from minor variations in experimental detail which may exert a profound influence on these sensitive *in vitro* culture conditions. EAE has been studied by Bornstein and Appel (1961), showing demyelination, and further studies by these workers have suggested that the agent responsible is a conventional antibody (Appel and Bornstein, 1964); the reaction is complement dependent. As in all systems requiring a visual assessment these tests are difficult to quantitate, and most workers use a simple score of three or four grades. A cytotoxic index has been used in a study with serum from monkeys with EAE (Lamoureux, Boulay and Borduas, 1966), but this has been criticized by others on a number of technical grounds (Hughes and Field, 1967). Experiments with sera from animals with 'allergic' disorders in organs other than the brain do not cause demyelination in culture. In multiple sclerosis up to 84 per cent of patients show demyelinative activity, whereas in normal controls figures up to 24 per cent have been reported. Activity is also present in 60 per cent of cases of motor neurone disease and 50 per cent in general paralysis of the insane. It must therefore be concluded that myelotoxicity of serum, while greatly increased in acute multiple sclerosis, is not absolutely specific to this condition. Similar tests using spinal fluid have produced one positive result in EAE (Bornstein and Appel, 1965) in the one sample tested, and Hughes and Field (1967) obtained only negative results in six unconcentrated spinal fluids from patients with multiple sclerosis. However, Lumsden (1966) reported three out of five spinal fluids as active following concentration to give globulin levels comparable with those in serum. Although it has been shown that the demyelinating activity resides in a globulin and that the reaction is complement dependent, it has not yet been proved to be antibody in nature.

It may be relevant to consider the fluorescent antibody studies of Rauch and Raffel (1964) which showed that antibody prepared against the basic protein of Roboz-Einstein reacted specifically with myelinated fibres in the central nervous system, and this very carefully controlled study points to the presence of antibody against a component of myelin, the accepted source of basic protein.

The position of platelet adhesiveness and other possible abnormalities in platelet behaviour is still somewhat anomalous. There have been no studies in EAE on technical grounds and there is general agreement that adhesiveness is increased in multiple sclerosis (Nathanson and Savitsky, 1952; Wright, Thompson and Zilkha, 1966; Millac, 1966; and others). One is forced, on the grounds of general non-specificity, to say that this may be an epiphenomenon in spite of the inherent attraction of a micro-thrombotic pathogenesis of multiple sclerosis. The use of the release of 5-hydroxytryptamine from platelets in the presence of an antigen–antibody reaction (Humphrey and Jaques, 1955) as a means of detecting antibodies to encephalitogen showed that antibody was present not only in most patients with multiple sclerosis but also in 20–30 per cent of normal individuals and in a number of cases suffering from cerebral tumours (75 per cent) (Caspary and Field, 1967).

EAE may be inhibited by priming the experimental animal with en-cephalitogen without adjuvant or, though less effectively, with Freund's complete adjuvant alone (Alvord, Shaw, Hruby and Kies, 1965; Field and Caspary, 1964; Cunningham and Field, 1965) before active challenge. This effect is permanent and also seen in animals allowed to recover from EAE and rechallenged. Antibody production is increased by this process (Caspary, Sinden and Field, 1966), but no other evidence on the nature of this inhibition has emerged. It would almost appear as if a 'decision process' may differ on a basis of presentation of a potential auto-allergen. Patients with multiple sclerosis who were skin tested with encephalitogen, the nearest approach to the experimental situation described above, showed no improvement in their clinical condition.

Studies on the chemistry of the demyelinating disorders have included a large number of studies on lipids, in brain and in body fluids, and in the case of the latter may be of diagnostic help. Other examinations fall more properly into the realm of chemical pathology and are reviewed by Lumsden (1966) as are also numbers of studies of a more academic nature.

Throughout this review an attempt has been made to compare and where necessary contrast the proven 'allergic' encephalomyelitis with multiple sclerosis, the most common of the group of allergic disorders. It will be assumed what holds true for this condition may also be true for the other members of the family of demyelinating diseases. There is little doubt that at least for cases of rabies vaccine encephalitis the parallel is excellent even though their rarity precludes studies on any scale. On clinical grounds the encephalitides following vaccination and those arising from viral infection provide a closer approach to EAE as these tend not to have the recurring nature of the disease in multiple sclerosis. Although recurrence has been described in viral or post-infectious neurological disease this is not a salient feature of this group. The allergic unifying hypothesis of demyelinating disease, though capable of explaining many of the features of disease, makes little attempt to probe at the initiation of this pathological process, particularly in the case of multiple sclerosis. Autoimmune disturbance following viral infection may occur, either by breaking the integrity of the organ and thereby permitting potentially dangerous antigens to escape or even by a metabolic disturbance encouraging the production of myelin components slightly modified so that the reticuloendothelial system no longer recognizes them as 'self'. The antigen escape rate may be critical, and the release of large amounts, as in cerebral injury or following surgery, could be protective on direct analogy with immune paralysis. In the case of multiple sclerosis a repeat stimulus to coincide with exacerbation of the disease must be postulated and it has tentatively been suggested that this may be viral or perhaps better 'Slow Virus' in nature (Palsson, Pattison and Field, 1965; Field, 1966). Repeat or booster stimulation could provide an explanation for some of the differences between EAE and multiple sclerosis, antibodies change on repeated stimulation and even though there is broad general agreement between workers as to the nature of encephalitogen the stimulating antigen released in disease may differ in some essential features from that isolated in the laboratory. If the occurrence of an immune process is accepted, on the basis of pathology, antibody studies and myelinotoxicity, then it may now be essential to investigate more directly the influence of the presumed toxic antibodies or antibody-carrying cells on nervous tissues in culture to determine their effect on cellular synthesis and metabolism at the most fundamental level.

REFERENCES

ADAMS R.D. and KUBIK C.S. (1952) *Am. J. Med.* **13**, 510
AHRENGOT V. (1957) *Acta psychiat. neurol. scand.* **32**, 192

ALVORD E.C. (1966) *J. Neuropath. exp. Neurol.* **25**, 1

ALVORD E.C., SHAW C.-M., HRUBY S. and KIES M.W. (1965) *Ann. N.Y. Acad. Sci.* **122**, 333

APPEL S.H. and BORNSTEIN M.B. (1964) *J. exp. Med.* **119**, 303

ASKONAS B.A. and HUMPHREY J.H. (1948) *Biochem. J.* **68**, 252

BALL E.J. and CASPARY E.A. (1963) *Life Sciences* **10**, 737

BAUER H. and HEITMAN R. (1958) *Dt. Z. NervHeilk.* **178**, 47

BÖHME D., LEE J.M., SCHNEIDER H.A. and WACHSTEIN M. (1966) *J. Neuropath. exp. Neurol.* **25**, 311

BÖHME D., PAAL G., KERSTEN W. and KERSTEN H. (1963) *Nature (Lond.)* **197**, 609

BORNSTEIN M.B. and APPEL S.H. (1961) *J. Neuropath. exp. Neurol.* **20**, 141

BORNSTEIN M.B. and APPEL S.H. (1965) *Ann. N.Y. Acad. Sci.* **122**, 280

BORNSTEIN M.B. and MURRAY M.R. (1958) *J. biophys. biochem. Cytol.* **4**, 499

CASPARY E.A. (1965) *J. Neurol. Neurosurg. Psychiat.* **28**, 61

CASPARY E.A. (1966) *J. Neurol. Neurosurg. Psychiat.* **29**, 103

CASPARY E.A. and BALL E.J. (1962) *Brit. med. J.* **ii**, 1514

CASPARY E.A. and FIELD E.J. (1963) *Dt. Z. NervHeilk.* **184**, 478

CASPARY E.A. and FIELD E.J. (1965) *Ann. N.Y. Acad. Sci.* **122**, 182

CASPARY E.A. and FIELD E.J. (1967) *Dt. Z. NervHeilk.* **190**, 267

CASPARY E.A., FIELD E.J. and BALL E.J. (1964) *J. Neurol. Neurosurg. Psychiat.* **27**, 25

CASPARY E.A., FIELD E.J., MCLEOD I. and SMITH C. (1963) *Z. ImmunForsch.* **125**, 459

CASPARY E.A., SINDEN R.E. and FIELD E.J. (1966) *Z. ImmunForsch.* **130**, 454

CHEEVER F.S., DANIELS J.B., PAPPENHEIMER A.M. and BAILY O.T. (1949) *J. exp. Med.* **90**, 181

COOMBS R.R.A., COOMBS A.M. and INGRAM D.G. (1961) In *The Serology of Conglutination and its Relation to Disease.* Blackwell, Oxford

CUNNINGHAM V.R. and FIELD E.J. (1965) *Ann. N.Y. Acad. Sci.* **122**, 346

DENCKER S.J. and SWAHN B. (1961) *Lund Universitats Arsskrift N.F. Avd. Z.* **57**, No. 10

FARR R.S. (1958) *J. infect. Dis.* **103**, 239

FERRARO A. (1944) *Archs Neurol. Psychiat (Chicago)* **52**, 443

FIELD E.J. (1966) *Brit. med. J.* **2**, 564

FIELD E.J. (1966) *J. Roy. Coll. Phycns. (Lond.)* **1**, 56

FIELD E.J. and CASPARY E.A. (1964) *Nature (Lond.)* **201**, 936

FIELD E.J. and MILLER H.G. (1961) *Archs int. Pharmacodyn. Thép.* **134**, 76

FIELD E.J. and RAINE C.S. (1966) *Am. J. Path.* **49**, 537

FIELD E.J. and RIDLEY A. (1960) *Brit. med. J.* **ii**, 1053

FIELD E.J., CASPARY E.A. and BALL E.J. (1963) *Lancet* **ii**, 11

FREUND J. and MCDERMOTT K. (1942) *Proc. Soc. exp. Biol. Med.* **49**, 548

FRICK E. and SCHEID-SEYDEL L. (1958) *Klin. Wschr.* **36**, 857

GAJDUSEK D.C., GIBBS C.J. and ALPERS M. (1965) *Slow, Latent and Temperature Virus Infections.* NINDB, Monograph No. 2

GLANZMANN E. (1927) *Schweiz. med. Wschr.* **57**, 145

HARWIN S.M. and PATERSON P.Y. (1962) *Nature (Lond.)* **194**, 391

HILD W. (1957) *Z. Zellforsch. mikrosk. Anat.* **46**, 71

HONEGGER C.G., RITZEL G. and RIEDER H.P. (1964) *Z. ImmunForsch.* **126**, 49

HOYER L.W., CONDIE R.M. and GOOD R.A. (1960) *Proc. Soc. exp. Biol. Med.* **103**, 205

HUGHES D. and FIELD E.J. (1967) *Clin expl. Immunol.* **2**, 295

HUMPHREY J.H. and JAQUES R. (1955) *J. Physiol.* **128**, 9

JELLINGER K. and SEITELBERGER F. (1958) *Klin. Wschr.* **36**, 437

KABAT E.A., WOLF A. and BEZER A.E. (1947) *J. exp. Med.* **85**, 117

KIBLER R.F. (1965) *Ann. N.Y. Acad. Sci.* **122**, 469

KIBLER R.F. and BARNES A.E. (1962) *J. exp. Med.* **116**, 807

KIES M.W. (1965) *Ann. N.Y. Acad. Sci.* **122**, 161

KIES M.W. and ALVORD E.C. (1959) In '*Allergic*' *Encephalomyelitis*. Thomas, Springfield, Illinois

KIES M.W., GOLDSTEIN N.P., MURPHY J.B., ROBOZ E. and ALVORD E.C. (1957) *Neurology (Minneap.)* **7**, 175

KIES M.W., SHAW C.-M., FAHLBERG W.J. and ALVORD E.C. (1960) *Ann. Allergy* **18**, 849

KIES M.W., THOMPSON E.B. and ALVORD E.C. (1965) *Ann. N.Y. Acad. Sci.* **122**, 148

KUHN P. and STEINER G. (1917) *Medsche Klin.* **13**, 1007

KUHN P. and STEINER G. (1920) *Z. Hyg. InfectKrankh.* **90**, 417

KUWERT E., FIRNHABER W. and PETTE E. (1964) *Ann. N.Y. Acad. Sci.* **122**, 429

LAATSCH R.H., KIES M.W., GORDON S. and ALVORD E.C. (1962) *J. exp. Med.* **115**, 777

LAMOUREUX G., BOULAY G. and BORDUAS A.G. (1966) *Clin. expl. Immunol.* **1**, 307

LAMOUREUX G., CARNEGIE P.R., McPHERSON T.A. and JOHNSTON D. (1967) *Clin. expl. Immunol.* **2**, 601

LEE J.M. and SCHNEIDER H.A. (1962) *J. exp. Med.* **115**, 157

LEVINE S., WENK E.J. and HOENIG E.M. (1967) *Transplantation*, **5**, 534

LEVINE S. and WENK E.J. (1965) *Ann. N.Y. Acad. Sci.* **122**, 209

LUMSDEN C.E. (1955) In *Proceedings of the 2nd International Congress of Neuropathology*, London, p. 429. Excerpta Medica Foundation, Amsterdam

LUMSDEN C.E. (1965) In *Multiple Sclerosis, a Reappraisal*. Ed. McALPINE D., LUMSDEN C.E. and ACHESON A.D. Livingstone, London and Edinburgh

LUMSDEN C.E. (1966) In *Proceedings of the 5th International Congress of Neuropathology*, p. 231. Excerpta Medica Foundation, Amsterdam

LUMSDEN C.E., ROBERTSON D.M. and BLIGHT R. (1966) *J. Neurochem.* **13**, 127

MACKAY J.R. and BURNETT F.M. (1963) *Autoimmune Diseases*. Thomas, Springfield, Illinois

MACKAY J.R. and GAJDUSEK D.C. (1958) *A.M.A. Archs internal Med.* **101**, 30

MacPHERSON C.F.C. (1962) *Can. J. Biochem. Physiol.* **40**, 1811

MILLAC P. (1967) *Dt. Z. NervHeilk.* **191**, 74

MILLER H.G. and SCHAPIRA K. (1959) *Brit. med. J.* **i**, 737, 811

MORGAN I.M. (1947) *Am. J. Hyg.* **45**, 390

NAKAO A. and ROBOZ-EINSTEIN E. (1965) *Ann. N.Y. Acad. Sci.* **122**, 171

NATHANSON M. and SAVITSKY J.P. (1952) *Bull. N.Y. Acad. Med.* **28**, 462

O'GRADY F. (1956) *Brit. J. Tuber. Dis. Chest* **50**, 159

PALSSON P.A., PATTISON I.H. and FIELD E.J. (1965) *Slow, Latent and Temperate Virus Infections* NINDB, Monograph No. 2, p. 49

PAPADOPOULOS N.M., HESS W.C., O'DOHERTY D. and McLANE J.F. (1959) *Clin. Chem.* **5**, 569

PATERSON P.Y. (1960) *J. exp. Med.* **III**, 119

PATERSON P.Y. (1966) *Advance Immunol.* **5**, 131
PATERSON P.Y., HANSON M.A. and GERNER E.W. (1967) *Proc. Soc. exp. Biol. Med.*
    124, 928
PETTE H. (1942) *Die acut Entzündlichen Erkrankungen des Nervensystem.* Thieme,
    Stuttgart
PETTE H., MANNWEILER K. and PALACIOS O. (1961) *Dt. Z. NervHeilk.* **182**, 635
PUTNAM T.J. (1935) *Archs Neurol. Psychiat. (Chicago)* **33**, 929
RAUCH H.C. and RAFFEL S. (1964) *J. Immunol.* **92**, 452
RIBBERT H. (1882) *Virchows Arch. path. Anat. Physiol.* **90**, 243
RIDLEY A. (1963) *Z. ImmunForsch.* **125**, 173
RIEDER H.P. and WÜTHRICH R. (1962) *Klin. Wschr.* **40**, 1070
RITZEL G., WÜTHRICH R. and RIEDER H.P. (1963) *Schweiz. med. Wschr.* **93**, 1336
ROBERTS S.D. (1962) *Lancet* **i**, 164
ROBERTSON D.M., BLIGHT R. and LUMSDEN C.E. (1962) *Nature (Lond.)* **196**, 1005
ROBOZ-EINSTEIN E., ROBERTSON D.M., DI CAPRIO J.M. and MOORE W. (1962) *J.
    Neurochem.* **9**, 353
ROSENTHAL F.D. and SOOTHILL J.F. (1962) *J. Neurol. Neurosurg. Psychiat.* **25**, 177
ROSS J. (1960) *Dt. Z. NervHeilk.* **181**, 159
SHAW C.-M., ALVORD E.C., FAHLBERG W.J. and KIES M.W. (1962) *J. exp. Med.* **115**,
    169
SHAW C.-M., ALVORD E.C., KAKU J. and KIES M.W. (1965) *Ann. N.Y. Acad. Sci.*
    122, 318
SIGURDSSON B. and PALSSON P.A. (1958) *Brit. J. exp. Path.* **39**, 519
STAUFFER R.E. and WAKSMAN B.H. (1954) *Ann. N.Y. Acad. Sci.* **58**, 570
STEINER G. (1952) In *Proceedings of the 1st International Congress of Neuropathology,
    Rome*, Vol. 1, p. 193. Rosenberg & Sellier, Torino, Italia
STONE S.H. (1962) *Int. Archs Allergy appl. Immun.* **20**, 193
SWANK R.L. (1950) *Am. J. med. Sci.* **220**, 421
SWINTHINBANK J., SMITH H.V. and VOLLUM R.L. (1953) *J. Path. Bact.* **65**, 565
THOMSON J.D. and AUSTIN R.W. (1962) *Proc. Soc. exp. Biol. Med.* **111**, 121
UCHIMURA I. and SHIRAKI H. (1957) *J. Neuropath. exp. Neurol.* **16**, 139
VAN BOGAERT L. (1932) *Revue Neurol.* **2**, 1
WAKSMAN B.H. (1956) *J. infect. Dis.* **99**, 258
WRIGHT H.P., THOMPSON R.H.S. and ZILKHA K.J. (1966) *Lancet* **ii**, 1109
YOKOYAMA M., TRAMS E.G. and BRADY R.O. (1962) *Proc. Soc. exp. Biol. Med.* **111**,
    350

CHAPTER FIVE

PHENOTHIAZINES AND RELATED SUBSTANCES IN PSYCHIATRY AND NEUROLOGY. CLINICAL EFFECTS

W. A. LISHMAN

The phenothiazines have had a dramatic and varied history in the past 20 years which can rarely have been equalled among therapeutic substances. In the 1930s, unsubstituted phenothiazine found a limited use as a urinary antiseptic, as a vermifuge in animals and as an insecticide (Friedman and Everett, 1964). In the 1940s promethazine was introduced as an antihistaminic with sedative properties. Thence the synthesis of a series of derivatives led in the Rhône-Poulenc Laboratories in 1950 to chlorpromazine. From this point onwards, chlorpromazine itself and a number of later phenothiazine derivatives have attracted astonishing attention and found application in many fields of medicine. The phenothiazines now constitute one of the most commonly prescribed classes of drugs in medicine.

The first clinical applications were in anaesthesia. Chlorpromazine was shown to potentiate anaesthetic agents and to suppress adaptive responses in a way which allowed hypothermia to be applied in patients undergoing surgery or suffering from shock (Laborit and Hugenard 1951). As a premedication for anaesthesia chlorpromazine was found to impart a state of relaxed calm and apparent indifference to distracting stimuli, yet without producing the degree of somnolence seen with other sedative agents. This unique effect was something new in therapeutics, and was designated by a new term—'tranquillization'.

Two lines of application soon followed. The tranquillizing effect proved useful in the management of distressing and terminal physical illness, especially since it was coupled with potentiation of analgesic agents and was found to have antiemetic properties (Friend and Cummins, 1953). And in psychiatry, chlorpromazine was used in combination with other drugs for producing prolonged narcosis in a variety of mental illnesses (Deschamps and Cadoret, 1953). But chlorpromazine's entry into psychiatry was truly marked by a paper read before a meeting for the centenary of the Société Medico-Psychologique by Delay, Deniker and Harl in 1952. Here chlorpromazine was reported to control violent excitement, without producing confusion or persistent somnolence, in a small group of patients which included cases of mania and of schizophrenia. In this relatively modest communication we see the most decisive event in shaping the future applications of the phenothiazines. Soon afterwards they entered with explosive suddenness into the management of schizophrenia, and became the focus of unprecedented research both in the laboratory and in clinical application. It has been variously claimed that in psychiatry the phenothiazines have 'heralded the clinical revolution that has transformed treatment and medical attitudes' (Sargant and Slater, 1963), or that they are 'responsible for the essential progress of psychiatry during the last decade' (Haase and Janssen, 1965). Be that as it may, their value in psychiatry cannot be disputed, and a progressively more critical attitude to their action and application has so far not served appreciably to curtail their use.

PHARMACOLOGICAL ACTIONS

Before discussing clinical applications it may be useful to look briefly at the general pharmacological actions of chlorpromazine and its derivatives.

The pharmacological range of chlorpromazine is wide. It shows antiadrenaline, antiacetylcholine and antiserotonin actions. Its own antihistaminic properties are weak but those of other phenothiazines may be strong. It potentiates a large number of analgesic, hypnotic and anaesthetic agents.

In the central nervous system there is little demonstrable direct action upon the neocortex. Actions on the hypothalamus are shown by hypothermic and antipyretic effects, by depression of vasomotor reflexes, and perhaps by the antiemetic properties. Neurophysiological experiments clearly show a pronounced action on brain stem mechanisms. Chlorpromazine has very little effect on the arousal response to direct stimulation of the brain stem reticular formation (whether measured behaviourally or by EEG), but markedly raises the threshold for arousal by incoming stimuli along the ascending sensory pathways. This dissociation of action is not seen with barbiturates, and forms an interesting paradigm of the clinical 'tranquillizing' effect. It can be shown that, while raising the threshold for general arousal to auditory stimulation, there is no blocking of the arrival of stimuli at the auditory cortex itself; the brain stem action would therefore seem to be on *collateral* relays arising from the ascending sensory pathways and impinging on the reticular formation (see Bradley, 1963).

The descending reticular pathways are affected in animals (Henatsch and Ingvar, 1956) but less certainly in man, and there is no direct action on the spinal cord. The pronounced action on the basal ganglia and extrapyramidal system will be considered later. The EEG is transiently disturbed by intravenous chlorpromazine, showing the features normal to drowsiness (Brazier, 1964); in brain-damaged or epileptic subjects there is an intensification of disturbance with generalized dysrhythmia and focal slowing (Hollister, 1961).

Behaviourally a number of interesting effects on the CNS are seen in animals (Bradley, 1963). Chlorpromazine increases measures of sociability in cats and abolishes sham rage in the decorticate cat. It tames Rhesus monkeys and protects rats from the stress induced by change of environment. In many animals it can abolish a newly acquired conditioned avoidance response at a dosage level which leaves the unconditioned escape response intact (cf. barbiturates which block both responses together). In man, of course, the outstanding behavioural effect is sedation without impairment of consciousness; a large initial dose of chlorpromazine may impair cognitive function and psychomotor control in much the same way as do barbiturates, but with continued administration tolerance is acquired and the effect recedes (Kornetsky, Pettit, Wynne and Evarts, 1959). This parallels the clinical observation of some initial drowsiness which passes in spite of a steadily increasing dose, and immediately establishes the superiority of phenothiazines over other sedatives in the management of excitable and overactive patients.

UNWANTED EFFECTS

Phenothiazines appear to be remarkably safe drugs despite their many pharmacological actions. Successful suicide does not appear to have been recorded despite acute doses of up to 20 g of chlorpromazine (Hollister, 1961). No adverse effects have been found in children born to mothers taking up to 150 mg of chlorpromazine daily during pregnancy (Kris, 1962), though doses over 500 mg per day in the last weeks of pregnancy may depress the respiratory centre of the newborn (Sobel, 1960).

The complication of jaundice (approximately 0.5 per cent of cases), which appears to be an idiosyncratic effect, is usually benign and resolves rapidly on withdrawal of the drug. If it is to occur, it occurs during the early weeks of treatment, and is not related to dose. There is no evidence of subclinical hepatocellular damage in the absence of jaundice even on very prolonged administration (Cohen and Archer, 1955). Hypotension secondary to hypothalamic disturbance is fairly common in the early stages of treatment, and more rarely there are hormonal changes including feminization in men and menstrual irregularity and lactation in women. Toxic effects on the formation of white blood corpuscles in the bone marrow are very rare but occasionally occur.

Photosensitivity skin reactions and stellate lens opacities appear to be interrelated, and the latter seem to correlate with duration of therapy (Cairns, Capoore and Gregory, 1965; Barsa, Newton and Saunders, 1965). Progressive pigmentation of skin over exposed areas is a common sequel of prolonged phenothiazine treatment. Greiner and Nicolson (1965) have reported that pigment deposition is also increased in many organs of the body and that the distribution follows closely that of metabolites of phenothiazines. The role of melanin metabolism in schizophrenia has recently become the focus of new interest (see Smythies, 1967), especially in view of a reported benefit from treatment with penicillamine (Nicolson, Greiner, McFarlane and Baker, 1966).

Finally, one must note that phenothiazines have recently been incriminated in contributing to sudden unexplained death in a very small minority of patients (Hollister and Kosek, 1965). Examination of records of death in a large psychiatric hospital has shown an increase in unexpected deaths since the introduction of phenothiazines there in 1957 (Richardson, Graupner and Richardson, 1966). Intramyocardial lesions have been found at post-mortem, and acute ventricular fibrillation has been proposed as the mechanism of death.

The neurological changes which may follow prolonged administration of phenothiazines will be considered in detail later.

CHEMICAL STRUCTURE

A bewildering number of phenothiazines are currently in use. All share the common structure shown in Table 1 and vary principally in alterations to the radicals attached to C_2 and N. These may change the pharmacological properties considerably, and may even confer analgesic or antiparkinsonian effect (e.g. methotrimeprazine, ethopropazine). Most importantly they may profoundly alter their therapeutic potency, milligram for milligram, and perhaps also their efficacy in psychiatric disorder. Substitution of Cl for H at R_2 increases potency (i.e. promazine→chlorpromazine) and this is still further increased by the substitution of a CF_3 radical (triflupromazine). All three compounds, however retain their aliphatic side chain at R_1. The replacement of the latter by a piperidine ring or by a piperazine ring provides further series of compounds. The piperazine compounds show enormously increased potency, up to one hundred times that of chlorpromazine, milligram for milligram. In the table these compounds are arranged very approximately in order of increasing potency. Increased potency closely parallels an increased tendency to produce extrapyramidal side-effects, but is not, of course, necessarily related to increased usefulness in psychiatric disorder. This question will be considered in some detail later.

TABLE I

R RADICAL

Aliphatic	Piperidine	Piperazine
Promazine (Sparine)	Mepazine (Pacatal)	
Chlorpromazine (Largactil)	Thioridazine (Melleril)	
Triflupromazine (Vesprin)		
		Prochlorperazine (Stemetil)
		Perphenazine (Fentazin)
		Thiopropazate (Dartalan)
		Trifluperazine (Stelazine)
		Thioproperazine (Majeptil)
		Fluphenazine (Moditen)

It is worth noting in passing what a surprisingly small change in the phenothiazine nucleus is required to produce antidepressive effect, even though the phenothiazines themselves have negligible antidepressive effect. The replacement of the sulphur molecule by two methylene groups results in imipramine, perhaps the most clearly established of all anti-depressant drugs. Moreover, the drug 'opipramol', which may be regarded structurally as mid-way between the phenothiazines and imi-pramine, has been reported to show both tranquillizing and antidepressant effects (Kiloh, 1963), though here adequate trials do not yet appear to have been performed.

CLINICAL USE IN PSYCHIATRY

Following the report by Delay *et al* in 1952 the tranquillizing effect of chlorpromazine was soon exploited in mental hospitals throughout the world. Reports multiplied of its ability to calm excitement and aggression and to reduce overactivity among emotionally disturbed patients. An early controlled study by Elkes and Elkes (1955) was able to show convincingly that a true pharmacological action lay behind the enthusiastic reports of improved behaviour among chronic psychotic patients who posed difficult nursing problems. Chlorpromazine proved greatly superior to existing chemical remedies—barbiturates, paraldehyde and bromides—which all too often controlled excitement only at the expense of profound torpor and sleepiness. The rauwolfia alkaloids, introduced at about the same time, were quickly ousted by the phenothiazines on account of their tendency to produce depression and their relative slowness to take effect. Only the butyrophenones (haloperidol, triperidol), a group of drugs which are clinically similar while chemically distinct from the phenothiazines, have proved able in recent years to compete for clinical usefulness and popularity; the definitive place of the butyrophenones remains to be established, but to date they have only rarely been proposed as serious competitors with the phenothiazines.

The phenothiazines are often given credit for making 'open door policies' and the community care of the mentally ill realistic propositions. It is hard, of course, accurately to disentangle one cause from another in the steady revolution which has been occurring in the care of the mentally ill. We may note that statistics bearing on the efficacy of mental hospital treatment have sometimes indicated that improvement set in some time *before* the introduction of psychiatric drugs (Odegaard, 1964), so that the latter may perhaps more properly be regarded as adjuncts rather than prime movers in getting patients back into the community (Lewis, 1959). We may also note that the clinical state of chronic schizophrenics in different mental hospitals has been found to relate more closely to the social environment of each hospital rather than to differences in the drug dosages employed (Wing and Brown, 1961). Certainly the two factors—improvement in social policy and application of effective drugs—have proceeded hand in hand, and probably each has helped to realize the potentialities of the other.

Let us now consider more closely the mode and quality of improvement obtained with phenothiazines in different diagnostic categories of mental illness.

F

SCHIZOPHRENIA

Phenothiazines have come to form the mainstay of treatment in schizophrenia. In a recent questionnaire issued to senior psychiatrists in England it was found that 96 per cent of those who replied used phenothiazines, with or without ECT and other drugs, in the management of schizophrenic patients (Willis and Bannister, 1965). Indeed, a study of diagnosis–treatment relationships in psychiatry showed the highest concordance of all for this particular clinical practice (Bannister, Salmon and Leiberman, 1964).

An acute schizophrenic illness can, in a large proportion of cases, be brought under reasonable control within weeks of starting phenothiazine treatment. The more florid the clinical picture, the more obvious is the therapeutic effect. Less success is generally claimed in the withdrawn, quiet and inaccessible schizophrenic patient, though here, too, a measurable success is often claimed. In the N.I.M.H. (1964) multihospital controlled trial involving 344 acute schizophrenics, 95 per cent showed improvement with phenothiazines, and in 75 per cent this was judged marked to moderate in degree (cf. 50 per cent and 23 per cent for placebo). After 6 weeks 45 per cent of drug-treated patients were symptom free or showed only borderline illness. Ratings of symptoms and abnormal behaviour allowed a careful investigation of the specific areas of improvement, and showed that the phenothiazines were effective in controlling a very wide range of disturbances—thought disturbance, delusions, auditory hallucinations, ideas of persecution, social withdrawal, loss of self care and, of course, anxiety and agitation. An earlier study of acute schizophrenia gave essentially similar results (Casey, Lasky, Klett and Hollister, 1960).

In chronic schizophrenia, as might be expected, less incontrovertible evidence has been presented. The consensus of opinion would appear to be that phenothiazines can produce useful symptomatic improvement in a substantial proportion of cases, especially where there is prominent aggressiveness, overactivity or tension. With very high dosage sustained remission has been reported in up to 40 per cent of chronic schizophrenics in some series (Rosati, 1964). But in general the more rigorous the therapeutic trial the less are the claims put forward (Heilizer, 1960), and isolated reports continue to be made in which no benefit whatever can be discerned. This appears to be particularly true of the chronic apathetic states which may supervene after long continued schizophrenia (Letemendia and Harris, 1967). Where chlorpromazine has been evaluated against a setting of increased occupational and social activities, both are found to contribute to the improvement seen in chronic schizophrenics. Here it would appear that drugs have a greater effect on the speed than on the final level of improvement (Grygier and Waters, 1958). The balance of evidence suggests that the longer the duration of illness the more prominent are social and environmental factors, and the less decisive are drugs, in determining the degree of symptom relief and social adjustment (Cawley, 1967; Bennett, 1967).

A possible exception may lie in the chronic paranoid illnesses of later life. Post (1966) has reported that among such patients success or failure is highly significantly related to the adequacy of treatment with pheno-thiazines. Of seventy-one patients adequately treated, full remission was obtained in forty-three, marked improvement in twenty-two, and no improvement in six. These figures are impressive for a condition no-toriously resistant to other attempts at treatment; and in Dr Post's own series only one of twenty-two patients not on adequate doses of pheno-thiazines experienced a good remission.

The definitive value of phenothiazines on the long-range functioning of schizophrenic patients remains to be fully determined (Gittleman, Klein and Pollack, 1964). Attempts to assess the value of maintenance therapy have given conflicting results, not surprisingly in view of the wide range of clinical material, and of settings in which follow-up is attempted. Separate questions would seem to be posed by the recently acute schizo-phrenic in full remission, and the established chronic patient in whom symptoms need constantly to be kept in check. Several studies concur, however, in showing that relapse and rehospitalization is significantly more frequent on placebo than on long-term phenothiazine treatment, and that long-term social adjustment is improved by maintenance drug treatment (Diamond and Marks, 1960; Gross, Mitchman, Reeves, Lawrence and Newell, 1961; Katz and Cole, 1962; Pasamanik *et al*, 1964). Some evidence, however, suggests that relapse is merely delayed rather than prevented (Englehardt, Freedman, Rosen, Mann and Margolis, 1964) so that the final outcome may not be significantly altered.

DIFFERENCES BETWEEN DIFFERENT PHENOTHIAZINES
The great number of different phenothiazines currently available shows the vigour with which improvements over chlorpromazine have been sought. Indeed, rich rewards must accrue to the drug company which can produce a more effective or a less troublesome phenothiazine,

in view of the enormous numbers of patients on whom it might be expected to be used.

There are fashions in psychiatry, and some practitioners have strong personal preferences for individual phenothiazines. Experience has sometimes tended to indicate that certain drugs are especially useful in certain types of schizophrenia or on certain target symptoms of schizophrenia. But the literature here is vast and confused, and a critical review enables relatively few firm conclusions to be drawn. It seems clear that to the phenothiazines listed in Table 1, promazine and mepazine are inferior to the others in the treatment of schizophrenia (Casey, Lasky *et al*, 1960; Casey, Bennett *et al*, 1960; Kurland *et al*, 1961); and that among the others, piperazine compounds have a greatly increased tendency to produce extrapyramidal side-effects in a manner which fairly closely parallels their potency, milligram for milligram. But a further difference between aliphatic and piperazine compounds is often claimed, viz. that the latter have a lessened tendency to sedate and indeed a tendency to arouse and activate the patient (Sargant and Slater, 1963; Malitz, 1964). It is therefore sometimes suggested that piperazine compounds should be the drugs of choice for apathetic and underactive schizophrenic patients, and occasional careful studies have appeared to show that here piperazine compounds are indeed superior to chlorpromazine (Gwynne *et al*, 1962). In normal subjects, also, it has been confirmed by extensive psychological studies that piperazine compounds can at certain dose levels produce psychomotor stimulation (Di Mascio, Leston, Havens and Klerman, 1963). None the less, the majority of studies, including some of the most extensive and rigorous trials, have failed to demonstrate significant qualitative differences between piperazine compounds and chlorpromazine in the actual management of schizophrenia (Casey, Lasky *et al*, 1960; Adelson and Epstein, 1962; N.I.M.H., 1964). The repeated conclusion appears to be that chlorpromazine can still hold its own alongside the more recently introduced phenothiazine compounds. Nor have significant therapeutic differences emerged between different members of the piperazine class, provided they are properly tested one against the other according to a flexible dosage schedule. The failure to demonstrate group differences does not, of course, mean that individual susceptibility may not still vary; and it remains possible that with finer methods of evaluation qualitative differences among phenothiazines may yet become firmly established. In the meantime, therefore, clinical practice justifies the trial of different phenothiazines in the patient who responds poorly to the initial therapeutic endeavour.

PHENOTHIAZINES TRULY 'ANTIPSYCHOTIC' OR NOT?

The mechanism of action of phenothiazines remains, of course, very incompletely understood. This is not rare in therapeutic substances used in medicine. But a question which deserves careful consideration is whether they merely act symptomatically to damp down the outward manifestations of the schizophrenic illness (in a manner analogous to anticonvulsants in the treatment of idiopathic epilepsy), or whether they influence the disease process itself in some more fundamental way. Protagonists of the theory that phenothiazines possess a true 'antischizophrenic' action point to the rapidity and completeness of remission of the acute attack, and to the persistence of remission in some cases without maintenance drug therapy. In such cases the schizophrenic process is thought to have been halted by the action of the drug on unspecified pathology within the central nervous system. The conception is attractive but lacks pathological, biochemical or indeed epidemiological support. The completeness of recovery which can nowadays be seen in favourable circumstances cannot easily be divorced from improvements in hospital environment, nor the stability of recovery from current social policies and rehabilitation. It remains conceivable that drugs, merely by controlling key symptoms and key patterns of behaviour disturbance, can render the patient susceptible to a host of more subtle influences which bear very directly on his final recovery. Cawley (1967) closely argues a further point of relevance here: if phenothiazines truly influence the morbid process of schizophrenia in a specific way, one might expect to find that factors of known prognostic good or bad omen in the untreated case were not necessarily the same after exposure to phenothiazines. No evidence is yet to hand to support this proposition.

Careful studies have been made of the range of schizophrenic symptoms which improve with phenothiazine medication. An example has already been detailed above from the N.I.M.H. multihospital controlled study (N.I.M.H., 1964). It is to be noted especially that not only do anxiety and agitation respond, but also thought disturbance, delusions, hallucinations, ideas of persecution, etc. So-called negative symptoms such as withdrawal, inertia, underactivity, are also reported to improve, perhaps less certainly but in an impressive number of studies incontrovertibly. Arguing from this range and scope of effect, some are tempted again to postulate an 'antischizophrenic' action (N.I.M.H., 1964; Cole and Mattson, 1965). It would seem, however, incautious to draw firm conclusions from such data, because we know very little about the ways in which schizophrenic symptoms may depend one upon the other, and virtually nothing of the mechanisms by which individual symptoms are derived from any fundamental 'psychotic process'.

USE IN OTHER PSYCHIATRIC DISORDERS

Phenothiazines have found wide application in psychiatric disorders other than schizophrenia. Few conditions characterized by profound excitement, tension and overactivity have failed to benefit to some degree.

Early reports emphasized the value of chlorpromazine in manic and hypomanic patients (e.g. Lehmann and Hanrahan, 1954) and this is still widely regarded as the treatment of first choice. Butyrophenones such as haloperidol have tended recently to replace chlorpromazine in the hands of some clinicians, but their superiority has not been established. A high proportion of cases respond to chlorpromazine from the outset and the disorder is brought rapidly under control. If withdrawn too soon, however, symptoms are likely to return, suggesting that the disease process is not in any fundamental sense altered. The liability to spontaneous remissions makes the latter point extremely difficult to assess. Maintenance treatment is often prescribed for patients prone to recurrent attacks, but no firm evidence points to effectiveness in delaying or aborting further attacks.

In severe depressive illnesses the phenothiazines are often used to allay agitation and tension, but, unlike imipramine and amitriptyline, they have little or no effect on the depressive mood itself nor on the other associated symptoms of depression.

In dementia the clinical usefulness of chlorpromazine has often been stressed very highly (e.g. Baker, 1955). The tranquillizing effect on disturbed and restless senile patients may transform the problems of their day-to-day management. Paranoid symptoms may yield, or at least lose force and distressing intensity. Unlike barbiturates, phenothiazines in moderate dosage appear to be relatively safe in the elderly, and much less liable to further confuse the failing brain. Organic confusional states from a variety of toxic and infective causes have been shown to benefit (Meyer, 1956; Delay, Deniker and Ropert, 1959). Chlorpromazine is a mainstay of treatment in delirium tremens, and the decreased mortality from this disease has been attributed to it (Coirault *et al*, 1956). Post-operative confusional states and psychoses may rapidly be brought under control, and with the help of phenothiazines can often continue to be nursed in general hospital beds. Other applications are in profoundly disturbed institutionalized epileptics (Bonafede, 1955) and in alcoholic and narcotic withdrawal (Van Grasse, 1958; Friedgood and Ripstein, 1955). Addiction to phenothiazines themselves has not been reported, which greatly adds to the confidence with which they can be used in the latter circumstances. The value of phenothiazines in neurotic illness, however, appears to be small. Early reports that chlorpromazine could help anxiety and tension states, obsessional states and phobias (Winklemann, 1954) have not been confirmed by fully controlled investigations (Merry *et al*, 1957; Raymond *et al*, 1957). More recent drugs such as chlordiazepoxide have made more impact in this corner of psychiatry. But chlorpromazine would appear to retain a useful place in some treatment regimes for anorexia nervosa (Crisp, 1965), and along with barbiturates, it remains an integral part of treatment by continuous narcosis.

Thus, in psychiatric disorders other than schizophrenia, the sedative and calming action of phenothiazines is obviously of paramount importance; complex questions of any additional mechanisms of action are scarcely raised. And in disorders other than schizophrenia chlorpromazine has given way much less to trials of the newer phenothiazine derivatives.

PHENOTHIAZINES AND NEUROLOGY
Phenothiazines have found a limited place in the management of certain neurological disorders. An early report, for example, showed the value of chlorpromazine in the management of acute head injuries (Shea, Alman and Fazekas, 1955). The restless and disturbed behaviour commonly seen in the phase of post-traumatic confusion responds to chlorpromazine, and may enable the patient to be nursed in a quiet tranquil state while necessary diagnostic and therapeutic measures are carried out. The advantages over heavy barbiturate sedation are considerable—the patient may more easily be roused for neurological examination, and there is little added danger of central respiratory depression. Signs of existing or progressing neurological damage are not masked. For similar reasons, phenothiazines have found useful application in the management of subarachnoid haemorrhage and other acute neurological illnesses where safe sedation is essential.

The aid to analgesia is of course an important additional benefit. Quite distinct are the potentiation of analgesics and the capacity of phenothiazines to modify the patient's reaction to painful stimuli. The latter effect has been compared to the action of frontal leucotomy—pain is still felt, but intrudes less on conscious awareness and provokes less emotional and motor response. The patient appears almost indifferent to his pain, and speaks of it more objectively. Chlorpromazine has found application in cases of causalgia (Dabbs and Peirce, 1955) and in cases of intractable neuritic pain (Sadove *et al*, 1954). It has been reported to have remarkable effect in acute episodes of porphyria (Melby, Street and Watson, 1956); the relief of pain, apprehension and agitation may be dramatic, though any paralysis present remains unaltered. Unlike barbiturates, phenothiazines do not aggravate or provoke porphyric episodes. (They do, however, aggravate myasthenia gravis (McQuillen, Gross and Johns, 1963).)

Intravenous chlorpromazine has been reported to reduce spasticity due to a variety of causes (Basmajian and Szatmari, 1955; Rushworth, Lishman, Hughes and Oppenheimer, 1961). This appears to be a specific effect on the mechanisms regulating muscle tone—it can be seen in the absence of any induced drowsiness, and voluntary power is unaffected. Unfortunately, the effect is transitory and has not found clinical application. In tetanus, however, chlorpromazine proves useful in suppressing muscular spasms, presumably through an action on the brain stem which results in suppression of the gamma motor system (Laurence and Webster, 1963). Finally, chlorpromazine has been reported to inhibit choreic and athetotic movements and to improve co-ordinated habitual movements in a variety of neurological diseases (Walther-Buel, 1956; Haase and Janssen, 1965). Marked motor improvement has been noted in up to a third of patients with Huntington's chorea. Here it is of course hard to be sure that the effect is not merely secondary to the tranquillizing effect of the drug on emotions; but it has been suggested that choreic and parkinsonian disturbances are to some extent antagonistic to one another and that the choreics benefit from a minor degree of drug-induced parkinsonism.

EXTRAPYRAMIDAL EFFECTS

In comparison to psychiatry, phenothiazines have but limited use in clinical neurology. Their neurological interest is, however, profound, by virtue of their capacity to induce parkinsonism and other disturbances of the extrapyramidal system. By most this is regarded as their chief disadvantage in therapy, though as we shall see later this effect has sometimes been seen as the key to their actions in psychotic illness.

It is certainly a striking observation that the three classes of drugs most effective in treating the psychoses—the phenothiazines, the butyrophenones and the rauwolfia alkaloids—should share a capacity to disturb extrapyramidal function. There are very few other drugs which do so. And while exact comparisons are difficult, the broad generalization may be stated that increasing potency in controlling psychotic overactivity and excitement parallels very closely the liability to provoke extrapyramidal disturbance (Table 2).

An accurate classification is difficult and the terminology in the literature is somewhat confused. But the three main syndromes shown in Table 3 are usually distinguished.

The dystonic syndromes are relatively rare, occurring in some 2–3 per

TABLE 2

	Average daily dose Effective in schizophrenia	Incidence of extrapyramidal effects
Promazine	—	Nil
Chlorpromazine	300 mg	35%
Trifluperazine	15 mg	60%
Haloperidol	3–6 mg	80%

(Based on data of Ayd, 1961 and Friedman and Everett, 1964.)

TABLE 3

Extrapyramidal syndromes

	Sex	Age	Timing	Response to anti-Parkinson drugs
Dystonic syndromes	M > F	Children and young adults	Abrupt and early	Excellent
Parkinsonian syndrome	F > M	Middle aged and elderly	First weeks Dose related	Excellent
Akathisia and Takathisia	F > M	Middle aged	First weeks	Good
Dyskinetic Syndromes				
Facial Dyskinesias	F > M	Elderly brain damaged	Often after months–years	Sometimes irreversible

cent of cases (Ayd, 1961). They alone show a preponderance in males, and are commoner in children and young adults than in later life. Onset is characteristically abrupt, occurring within the first few days of treatment or even after a single large dose. There is spasmodic or sustained tonic contraction of muscle groups, chiefly affecting the head and neck, but sometimes also the limbs and trunk. A variety of syndromes are seen—torticollis, retrocollis, trismus, deviation of head and eyes, occulogyric crises, opisthotonous, torsion of the trunk and limbs. In severe cases protrusion of the tongue and stridorous breathing may even embarrass respiration. The clinical picture may closely simulate dystonia musculorum, and may sometimes be misdiagnosed as tetanus. Hysteria may be suspected, especially since crises may be precipitated by emotional stress and can sometimes respond to suggestion (Deniker, 1960). Untreated, the disturbance subsides within a few hours of discontinuing phenothiazines, and fortunately it responds rapidly and completely to antiparkinsonian drugs which may be given intravenously in an emergency. An isolated case is reported in the literature (Druckman, Seelinger and Thulin, 1962) of persistent dystonia—drawing back of the head. This was ultimately lessened by bilateral thalamotomies.

The parkinsonian syndrome is very much commoner. The figure of 15 per cent (Ayd, 1961) is almost certainly an underestimate if minor degrees of hypokinaesia are included. Parkinsonian changes may be seen at any age in adults, but increase in frequency as age advances. Individual susceptibility varies greatly and a genetic factor appears to operate here. An interesting study by Myrianthopoulos, Kurland and Kurland (1962) has shown a three-fold increase in the incidence of naturally occurring Parkinson's disease among relatives of patients who develop this complication on phenothiazines. Onset is closely related to dosage and a clear threshold effect can often be demonstrated. Some patients, however, fail to show parkinsonian symptoms whatever the dose (Hollister, 1964). The earliest changes are to be detected in the handwriting (Haase and Janssen, 1965) and progress to slowing of movement, muscular rigidity and the typical parkinsonian gait, posture and facial expression. A coarse rhythmic resting tremor, usually of a single limb, may accompany the changes in tone or may occur in isolation. Clinicians do not agree whether drug-induced and clinical parkinsonism are identical, but the similarity in the fully evolved case can be very close. The condition responds to lowering of the dose of phenothiazines or may be kept in abeyance by the use of antiparkinsonian drugs.

Dyskinetic syndromes have recently become the subject of some concern. Akathisia (inability to sit still) and takathisia (a constant tippling and shuffling of the feet) are very common (21 per cent of cases, Ayd, 1961). They commonly respond to antiparkinsonian drugs but may sometimes prove difficult to control. More serious are the facial movements distinguished by the name 'facio-bucco-linguo-masticatory dyskinesia'. This has been reported in up to 5 per cent of female patients chronically treated with phenothiazines (Hunter, Earl and Thornicroft, 1964). Onset may be late in treatment, often after many years, or may begin some months after a prolonged course has been discontinued. No particular drug or dose is specially incriminated and it may develop while on regimes previously well tolerated for long periods of time. The condition consists of continuous grimacing, blinking, munching, sucking movements, with protrusions of the tongue, bulging of the cheeks and writhing movements of the jaw. Speech and eating accentuate the disability. The relationship to prolonged phenothiazine medication would now appear to be well established, but most cases have hitherto presumably been ascribed to schizophrenic mannerisms. When accompanied by dyskinetic movements of the limbs it may be mistaken for Huntington's chorea. It is much commoner in patients with evidence of brain damage, after leucotomy and after ECT (Faurbye *et al*, 1964), though such antecedents do not appear to be universal (Schmidt and Jarcho, 1966). Some cases respond to stopping the drug and some to antiparkinsonian medication, but a disconcerting number appear to persist indefinitely (Uhrbrand and Faurbye, 1960; Hunter *et al*, 1964; Schmidt and Jarcho, 1966). The resemblance to hyperkinetic encephalitic syndromes lead Hunter *et al*. (1964) to suggest that the patients suffer from a drug-induced encephalitic process. Autopsy information is unfortunately not yet reported. Whatever the cause, it seems that this distressing syndrome constitutes a small but definite risk which should be borne seriously in mind when prescribing long-term phenothiazine medication.

The mechanism of the extrapyramidal effects shown by phenothiazines is unclear. Post-mortem data on patients treated chronically with phenothiazines are scarce, but so far have failed to highlight basal ganglia lesions (Roizin, True and Knight, 1959). It has been hypothesized (Faurbye *et al*, 1964) that acute dystonic reactions result from altered permeability of neural cell membranes, and that parkinsonian rigidity may be due to disturbance of the function of dopamine in the caudate nucleus. Especial interest has attached to these possible mechanisms because a substantial body of opinion has considered that there is a close relationship between the eliciting of extrapyramidal effects and a therapeutic response in schizophrenia. This has led some clinicians to prescribe massive intermittent doses of phenothiazines in schizophrenia with the aim of producing short-lived episodes of severe extrapyramidal disturbance (Denham and

Carrick, 1960, 1961). An analogy is drawn with other psychiatric therapies in which a somatic disorder has been thought necessary in order to relieve mental illness, e.g. insulin coma therapy and electroconvulsive therapy (Denniker, 1960). Proponents of the value of extrapyramidal symptoms appear to refer principally to muscular hypertonus (Denham and Carrick, 1961). Haase and Janssen (1965) claim that fine extrapyramidal hypokinesia (detected by handwriting tests) is a *sine qua non* of 'antipsychotic action'.

These contentions have not, however, found wide support. It is hard to find convincing evidence that the concordance between extrapyramidal effect and therapeutic response is anything more than a simple dosage-response phenomenon. Nor is there good evidence that antiparkinsonian drugs given along with phenothiazines impair their therapeutic action. Opponents of the theory point to the widely attested therapeutic value of phenothiazines given in such a way, and with such antiparkinsonian cover, that extrapyramidal signs are not allowed to appear. Moreover, thioridazine appears to be as effective as chlorpromazine in both acute and chronic schizophrenia (Svendson, Faurbye and Kristjansen, 1961; Waldorp, Robertson and Voulekia, 1961; Stebanau and Grinols, 1964) yet appears to be remarkably less likely than most phenothiazines to produce extrapyramidal effects (Sandison, Whitelaw and Currie, 1960; Cole and Clyde, 1961). Finally, the large controlled trials which have failed to find differences in therapeutic efficiency between several phenothiazines (Casey, Lasky et al, 1960; N.I.M.H., 1964) have regularly shown marked differences in the incidence of extrapyramidal effects in the course of the trials.

In brief, it may be concluded with Shepherd, Rodnight and Lader (1967) that, while all effective tranquillizers appear capable of inducing extrapyramidal effects if given in big enough doses, these do not seem on present evidence to be essential for therapeutic response in the individual patient. The idea is less widely held now than formerly, and has made little impact on current therapeutic practices. It is, however, a possibility of great heuristic interest, and which still deserves definitive clarification in one direction or the other. It is also a possibility which has for a time drawn neurology and psychiatry somewhat closer together and united them in some common aims and interests.

ACKNOWLEDGMENT

I am greatly indebted to Dr Malcolm Lader for generous help in guiding me to the relevant literature.

REFERENCES

ADELSON D. and EPSTEIN L.J. (1962) *J. nerv. ment. Dis.* **134**, 543

AYD F.J. (1961) *J. Am. med. Ass.* **175**, 1054

BAKER A.A. (1955) *J. ment. Sci.* **101**, 175

BANNISTER D., SALMON P. and LEIBERMAN D.M. (1964) *Br. J. Psychiat.* **110**, 726

BARSA J.A., NEWTON J.C. and SAUNDERS J.C. (1965) *J. Am. med. Ass.* **193**, 10

BASMAJIAN J.V. and SZATMARI A. (1955) *A.M.A. Archs Neurol. Psychiatry* **73**, 224

BENNETT D. (1967) *Hospital Medicine* **1**, 589

BONAFEDE V.I. (1955) *A.M.A. Archs Neurol. Psychiatry* **74**, 158

BRADLEY P.B. (1963) In *Physiological Pharmacology*, Vol. 1. Ed. ROOT W.S. and HOFMAN F.G. Academic Press, New York and London

BRAXIER M.A.B. (1964) *Clin. Pharmac. Ther.* **5**, 102

CAIRNS R.J., CAPOORE H.S. and GREGORY I.D.R. (1965) *Lancet* **i**, 239

CASEY J.F., BENNETT I.F., LINDLEY C.T., HOLLISTER L.E., GORDON M.H. and SPRINGER N.N. (1960) *Archs gen. Psychiat.* **2**, 210

CASEY J.F., LASKY J.J., KLETT C.J. and HOLLISTER L.E. (1960) *Am. J. Psychiat.* **117**, 97

CAWLEY R.H. (1967) In *Recent Developments in Schizophrenia*. Ed. COPPEN A. and WALK A. *Br. J. Psychiat.*, special publication No. 1. Headley Brothers, Ashford, Kent

COHEN I.M. and ARCHER J.D. (1955) *J. Am. med. Ass.* **159**, 99

COIRAULT R., LABORIT H., MISSENARD R., JOLIVET B., HAINAULT J. and WEBER B. (1956) *Encéphale* **45**, 762

COLE J.O. and CLYDE D.J. (1961) *Revue can. Biol.* **20**, 565

COLE J.O. and MATTSON N. (1965) In *Neuropsychopharmacology*, Vol. 4. Ed. BENTE D. and BRADLEY P.B. Elsevier Publishing Co., Amsterdam

CRISP A.H. (1965) *Br. J. Psychiat.* **112**, 505

DABBS C.H. and PEIRCE E.C. (1955) *J. Am. med. Ass.* **159**, 1626

DELAY J. DENIKER P. and HARL J.M. (1952) *Annls méd.-psychol.* **110**, 112

DELAY J., DENIKER P. and ROPERT P. (1959) In *Psychopharmacology Frontiers*. Ed. KLINE N.S. Little, Brown & Co., Boston

DENHAM J. and CARRICK D.J.E.L. (1960) *Am. J. Psychiatry* **116**, 927

DENHAM J. and CARRICK D.J.E.L. (1961) *J. ment. Sci.* **107**, 326

DENIKER P. (1960) *Comp. Psychiat.* **1**, 92

DESCHAMPS A. and CADORET M. (1953) *Presse méd.* **61** (i), 878

DIAMOND L.S. and MARKS J.B. (1960) *J. nerv. ment. Dis.* **131**, 247

DiMASCIO A., LESTON M.A., HAVENS L. and KLERMAN G.L. (1963) *J. nerv. ment. Dis.* **136**, 15 and 168

DRUCKMAN R., SEELINGER D. and THULIN B. (1962) *J. nerv. ment. Dis.* **135**, 69

ELKES J. and ELKES C. (1954) *Brit. med. J.* **2**, 560

ENGLEHARDT D.M., FREEDMAN N., ROSEN B., MANN D. and MARGOLIS R. (1964) *Archs gen. Psychiat.* **11**, 162

FAURBYE A., RASCH P.-J., PETERSEN P.B., BRANDBORG G. and PAKKENBERG H. (1964) *Acta psychiat. scand.* **40**, 10

FRIEDGOOD C.E. and RIPSTEIN C.B. (1955) *New Engl. J. Med.* **252**, 230

FRIEDMAN A.H. and EVERETT G.M. (1964) *Adv. Pharmacol.* **3**, 83

FRIEND D.G. and CUMMINS J.F. (1953) *J. Am. med. Ass.* **153**, 480

GITTLEMAN R.K., KLEIN D.F. and POLLACK M. (1964) *Psychopharmacologia* **5**, 317
GREINER A.C. and NICOLSON G.A. (1965) *Lancet* **ii**, 1165
GROSS M., MITCHMAN I.L., REEVES W.P., LAWRENCE J.L. and NEWELL P.C. (1961) *Recent Adv. biol. Psychiat.* **3**, 44
GRYGIER P. and WATERS M.A. (1958) *A.M.A. Archs Neurol. Psychiatry* **79**, 697
GWYNNE P.H., HUNDZIAK M., KAVTSCHITSCH J., LEFTON M. and PASAMANICK B. (1962) *J. nerv. ment. Dis.* **134**, 451
HAASE H.-J. and JANSSEN P.A.J. (1965) In *The Action of Neuroleptic Drugs—a Psychiatric, Neurologic and Pharmacological Investigation.* North Holland Publishing Company, Amsterdam
HEILIZER F. (1960) *J. chron. Dis.* **11**, 102
HENATSCH H.-D. and INGVAR D.H. (1956) *Arch. Psychiat. NervKrankh.* **195**, 77
HOLLISTER L.E. (1961) *New Engl. J. Med.* **264**, 291
HOLLISTER L.E. (1964) *J. Am. med. Ass.* **189**, 311
HOLLISTER L.E. and KOSEK J.C. (1965) *J. Am. med. Ass.* **192**, 1035
HUNTER R., EARL C.J. and THORNICROFT S. (1964) *Proc. R. Soc. Med.* **57**, 758
KATZ M.M. and COLE J.O. (1962) *Archs gen. Psychiat.* **7**, 345
KILOH L.G., ROY J.R. and CARNEY M.W.P. (1963) *J. Neuropsychiat.* **5**, 18
KORNETSKY C., PETTIT M., WYNNE R. and EVARTS E.V. (1959) *J. ment. Sci.* **105**, 190
KRIS E.B. (1962) *Recent Adv. biol. Psychiat.* **4**, 180
KURLAND A.A., HANLON T.E., TATOM M.H., OTA K.Y. and SIMOPOULOS A.M. (1961) *J. nerv. ment. Dis.* **133**, 1
LABORIT H. and HUGENARD P. (1951) *Presse méd.* **59** (ii), 1329
LAURENCE D.R. and WEBSTER R.A. (1963) *Clin. Pharmac. Ther.* **4**, 36
LEHMANN H.E. and HANRAHAN G.E. (1954) *A.M.A. Archs Neurol. Psychiat.* **71**, 227
LETEMENDIA F.J.J. and HARRIS A.D. (1967) *Br. J. Psychiat.* **113**, 950
LEWIS A.J. (1959) In *Neuropsychopharmacology*, Vol. 1. Ed. BRADLEY P.B., DENIKER P. and RADOUCO-THOMAS C. Elsevier, Amsterdam
McQUILLEN M.P., GROSS M. and JOHNS R.J. (1963) *Archs Neurol. (Chicago)* **8**, 286
MALITZ S. (1964) In *Schizophrenia*, Vol. 1, No. 4. Ed. KOLB L.C., KALLMANN F.J. and POLATIN P. International Psychiatry Clinics. Little, Brown & Co., Boston
MELBY J.C., STREET J.P. and WATSON C.J. (1956) *J. Am. med. Ass.* **162**, 174
MERRY J., PARGITER R.A. and MUNRO H. (1957) *Am. J. Psychiatry* **113**, 988
MEYER H.-H. (1956) *Encéphale* **45**, 524
MYRIANTHOPOULOS N.C., KURLAND A.A. and KURLAND L.T. (1962) *Archs Neurol. (Chicago)* **6**, 6
N.I.M.H., PSYCHOPHARMACOLOGY SERVICE CENTRE COLLABORATIVE STUDY GROUP (1964) *Archs gen. Psychiat.* **10**, 246
NICOLSON G.A., GREINER A.C., McFARLANE W.J.G. and BAKER R.A. (1966) *Lancet* **i**, 344
ODEGAARD O. (1964) *Am. J. Psychiatry* **120**, 772
PASAMANICK B., SCARPITTI F.R., LEFTON M., DINITZ S., WERNERT J.J. and McPHEETERS H. (1964) *J. Am. med. Ass.* **187**, 177
POST F. (1966) In *Persistent Persecutory States of the Elderly.* Pergamon Press, Oxford and London
RAYMOND M.J., LUCAS C.J., BEESLEY M.L., O'CONNELL B.A. and FRASER ROBERTS J.A. (1957) *Brit. med. J.* **2**, 63

RICHARDSON H.L., GRAUPNER K.I. and RICHARDSON M.E. (1966) *J. Am. med. Ass.* **195,** 254

ROIZIN L., TRUE C. and KNIGHT M. (1959) *Res. Publs Ass. Res. nerv. ment. Dis.* **37,** 285

ROSATI D. (1964) *Br. J. Psychiat.* **110,** 61

RUSHWORTH G., LISHMAN W.A., HUGHES J.T. and OPPENHEIMER D.R. (1961) *J. Neurol. Neurosurg. Psychiat.* **24,** 132

SADOVE M.S., LEVIN M.J., ROSE R.F., SCHWARTZ L. and WITT F.W. (1954) *J. Am. med. Ass.* **155,** 626

SANDISON R.A., WHITELAW E. and CURRIE J.D.C. (1960) *J. ment. Sci.* **106,** 732

SARGANT W. and SLATER E. (1963) In *An Introduction to Physical Methods of Treatment in Psychiatry.* Livingstone, Edinburgh

SCHMIDT W.R. and JARCHO L.W. (1966) *Archs Neurol. (Chicago)* **14,** 369

SHEA J.G., ALMAN R.W. and FAZEKAS J.F. (1955) *A.M.A. Archs internal Med.* **96,** 168

SHEPHERD M., RODNIGHT R. W. and LADER M.H. (1967) In *Clinical Psychopharmacology.* The English Universities Press Ltd, London

SMYTHIES J.R. (1967) In *Recent Developments in Schizophrenia.* Ed. COPPEN A. and WALK A. *Br. J. Psychiat.*, special publication No. 1 Headley Brothers, Ashford, Kent

SOBEL D.E. (1960) *Archs gen. Psychiat.* **2,** 606

STABENAU J.R. and GRINOLS D.R. (1964) *Psychiat. Q.* **38,** 42

SVENDSEN B.B., FAURBYE A. and KRISTJANSEN P. (1961) *Psychopharmacologia* **2,** 446

UHRBRAND L. and FAURBYE A. (1960) *Psychopharmacologia* **1,** 408

VAN GASSE J.J. (1958) *Clin. Med. Surg.* **5,** 177

WALDORP F.N., ROBERTSON R.H. and VOURLEKIS A. (1961) *Comp. Psychiat.* **2,** 96

WALTHER-BUEL H. (1956) *Encéphale,* **45,** 771

WILLIS J.H. and BANNISTER D. (1965) *Br. J. Psychiat.* **111,** 1165

WING J.K. and BROWN G.W. (1961) *J. ment. Sci.* **107,** 847

WINKLEMAN N.W. (1954) *J. Am. med. Ass.* **155** (i), 18

CHAPTER SIX

BIOCHEMICAL PHARMACOLOGY OF PHENOTHIAZINES AND RELATED SUBSTANCES

G. Curzon

At a first glance the literature on the phenothiazines and related substances is of daunting complexity. There are getting on for a hundred such compounds on the market and a great range of biochemical and pharmacological activities have been reported to be affected by some of them. It may thus be useful to consider the significance of a few general properties. Drug action is a function of the shape and size of the drug molecule and of the distribution of electrons in it. Upon these factors depends whether the drug can get to its site of action, and how it interacts there. The structural elements associated with tranquillizing and antidepressant activity of these drugs will therefore be considered first.

STRUCTURE–ACTIVITY RELATIONSHIPS

Phenothiazines and related compounds have structures of the type shown in Fig. 1 and may be described in general as tricyclics. Phenothiazine

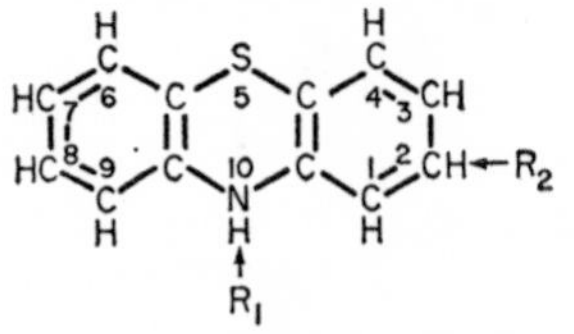

FIG. 1. General formula of phenothiazines.

itself is relatively inactive biologically, though it is used in veterinary medicine as an anthelmintic. However, by replacing a hydrogen atom on the nitrogen atom by various other groups (R_1), a series of compounds may be obtained which are pharmacologically and therapeutically active.

Abbreviations. DA = dopamine; NA = noradrenaline; 5HT = 5-hydroxytryptamine (serotonin).

Trade names. Largactil = chlorpromazine; Tofranil = imipramine.

Furthermore, by substituting a second group (R_2) into one of the outer rings, these activities may be modified. Thus from the parent molecule a series of active compounds may be obtained. If other atoms or groups are substituted for one or both of the sulphur and nitrogen atoms in the central ring of phenothiazine then other parent molecules are obtained from which other series of compounds may similarly be derived by substituting various R_1 and R_2 groups. Derivatives of the first parent substance in Table 1, phenoxazine, have little tranquillizing or anti-depressive activity (Ribbentrop and Schaumann, 1964) and derivatives of

TABLE I

Phenothiazine and other tricyclic molecules

	X^*	Y^*
Phenoxazine	O	NH
Phenothiazine ⎫ Azaphenothiazine† ⎬	S	NH
Thiaxanthene	S	=C—H
Iminodibenzyl	CH_2CH_2	NH
Iminostilbene	CH=CH	NH
Dibenzocycloheptadiene	CH_2CH_2	=C—H

* X and Y refer respectively to the atoms or groups which replace the ring S and NH of the phenothiazine molecule to give the above parent substances.
† In azaphenothiazines the ring CH at position 2 of phenothiazines is replaced by N.

the tricyclic hydrocarbon phenanthrene are also inactive. Derivatives of the other parent substances in Table 1 are active. Parent substances of the above inactive series have been shown by X-ray crystallography to be very different in shape from those of therapeutically active tricyclics (Hosoya, 1963). While phenanthrene and phenoxazine have planar structures, the latter are folded along the XY axis.

Effects on tranquillizing activity of altering R_1 and R_2 are shown in Table 2 which gives structures of some representative drugs. Those with marked tranquillizing activity have an R_1 substituent with a chain of three methylene groups of which one may be within an alicyclic ring (as in thioridazine). Activity is preserved in trimeprazine in which one of the methylene groups is substituted but is lost in derivatives with only two

G

methylene groups (promethazine, isothipendyl). At the end of the chain distal to the ring substituent there must also be a trisubstituted nitrogen atom which may be incorporated into an alicyclic ring (e.g. trifluperazine, thioridazine) or in the form $-N(CH_3)_2$. Demethylation results in decreased activity (Brune, Kohl, Steiner and Himwich, 1963). Although replacement of hydrogen at the 2 position of phenothiazines by the group R_2 is not essential for tranquillizing activity, it has been found that

TABLE 2

The effect on tranquillizing activity of changing R_1 and R_2

Drug	R_1	R_2	Tranquillizing activity
Phenothiazines			
Promazine	⎫	H	+
Chlorpromazine	⎬ $CH_2CH_2CH_2\ N(CH_3)_2$	Cl	+
Triflupromazine	⎭	CF_3	+
Trifluperazine	$CH_2CH_2CH_2\ N\overbrace{\quad}N(CH_3)_2$	CF_3	+
Trimeprazine	$CH_2CH\ CH_2N(CH_3)_2$ $\quad\ \ CH_3$	H	+
Thioridazine	$CH_2CH_2\langle\ \rangle$ $\qquad\quad N$ $\qquad\quad CH_3$	SCH_3	+
Methdilazine	$CH_2\underline{\quad}NCH_3$	H	—
Promethazine	$CH_2CH\ N(CH_3)_2$ $\quad\ \ CH_3$	H	—
	$COO\ CH_2CH_2N(CH_3)_2$	H	—
Azaphenothiazines			
Prothipendyl	$CH_2CH_2CH_2\ N(CH_3)_2$	H	+
Isothipendyl	$CH_2CH\ N(CH_3)_2$ $\quad\ \ CH_3$	H	—
Thiaxanthene			
Chlorprothixene	$=CH\ CH_2CH_2CH_2\ NC(H_3)_2$	Cl	+

chlorine, trifluoromethyl or other groups at this position enhance activity. In general, groups in the R_2 position which increase therapeutic activity markedly are electronegative and thus increase electron density on the rings. The requirement for $-C-C-C-N$ type R_1 side chain is reminiscent of the $-C-C-N$ type side chain of catecholamines, 5HT and histamine with which the tricyclic drugs have various pharmacological and biochemical relationships. Although deviations from the $-C-C-C-N$

chain structure result in compounds without tranquillizing properties many of them have antispasmodic, antimotion sickness, antitussive or antipruritic activity (Gordon, Craig and Zirkle, 1964).

The major tricyclic antidepressants are derivatives of aminodibenzyl and dibenzocycloheptadiene and are shown in Table 3. Drugs are commonly evaluated as potential antidepressive agents by their production of excitation in rats previously treated with reserpine or another amine releaser (Bickel and Brodie, 1964). Using this method it was found that reversal of reserpine effects by imipramine did not correlate with imipramine levels in the brain (Gillette, Dingell, Sulser, Kuntzman and Brodie, 1961). A metabolite, desipramine, formed by the removal of one methyl group from the terminal dimethylated nitrogen of imipramine was responsible. Desipramine was later found to be more effective as an

TABLE 3

Structures of antidepressive drugs

Drug	R_1	R_2
Iminodibenzyls		
Imipramine	$CH_2CH_2CH_2\,N(CH_3)_2$	H
Desipramine	$CH_2CH_2CH_2NH\,CH_3$	H
Dibenzocycloheptadienes		
Amitryptyline	$=CH\;CH_2CH_2N(CH_3)_2$	H
Nortryptyline	$=CH\;CH_2CH_2NH\,CH_3$	H

antidepressant than imipramine which, probably because of a weak direct phenothiazine-like tranquillizing action, may delay the antidepressive effect of its metabolite (Sulser, Watts, and Brodie, 1961). Similarly, the antireserpine effects of amitryptyline are due to the corresponding de-methylated metabolite nortryptyline (Bickel and Brodie, 1964). These authors applied the excitation test to a large number of drugs structurally related to imipramine. It was found that potential antidepressive activity was associated with an R_1 chain of the C—C—N or C—C—C—N type. Compounds with branched chains and longer chains were inactive. It appears that a rigid ring structure is not necessary as some activity remains when the central carbon bridge is broken. The two benzene rings are however essential. The most important point to come out of these structure–activity studies, however, is that —NH_2 or —$NH\,CH_3$ at the end of the R_1 chain is associated with antidepressive activity while —$N(CH_3)_2$ is associated with tranquillizing activity. A few substances with the latter group were active as antidepressants but this was probably due to their rapid demethylation. Demethylated derivatives of the more powerfully tranquillizing phenothiazines, for example desmethyltri-flupromazine, may have both tranquillizing and antidepressive activity and were suggested to be of potential value in the treatment of depressive states associated with anxiety and tension.

BINDING TO MEMBRANES

The tricyclic drugs have structures which are predominantly without separated positive or negative electrical charges, i.e. they are non-polar in character (except for their positively charged side amino nitrogen). This large non-polar component results in affinity for other substances which are also largely non-polar. Thus phenothiazines bind to lipids including those present as components of cell membranes. Chlorpromazine binds strongly to monomolecular films made from cholesterol esters (Zografi and Auslander, 1965) or from gangliosides (van Deenen and Semel, 1964). These films may be considered as models of biological membranes. When drugs bind to the latter they must affect membrane structure and may therefore be expected either to strengthen or weaken membranes or alter their permeability. Guth, Amaro, Sellinger and Elmer (1965) find that chlorpromazine stabilizes liver lysosome membranes and suggest that many reported inhibitions of enzymes by phenothiazines in membrane-containing systems such as homogenates or tissue slices may not be true inhibitions but simply due to the drug preventing release of bound enzymes into the assay medium. Decrease of membrane permeability by pheno-thiazines is well known and there are many reports of the hindering by phenothiazines of the transport of substances into various cells (reviewed by Guth and Spirtes, 1964). Chloropromazine inhibits the uptake of sodium, potassium, sulphate, methionine (Christensen, Feng, Polley and Wase, 1958) and acetate (Albaum and Milch, 1962) into brain and the incorporation of aminoacids into brain protein (Zoller, Schreier and Yang, 1958; Glasky, 1963). A related observation of some practical significance is that the intestinal absorption of phenothiazines themselves may be retarded by the previous administration of phenothiazines (Phillips and Miya, 1962).

Not only stabilization but also disruption of membranes by tricyclics has been reported. For example, erythrocytes are stabilized by low concentrations (Schales, 1953) but are weakened by higher concentrations of phenothiazines. The effects of a large series of tricyclic drugs on platelet and erythrocyte disruption and on surface activity have been compared by Ahtee (1966a, b). Platelet disruption was measured by 5HT release. Surface activity is a consequence of a molecule having polar groups (in this case the amino nitrogen) which give it affinity for water and also a nonpolar part which tends to pull it away from water, i.e. to the air-water interface where it alters surface tension. Surface activity depends on some of the factors which influence binding to lipid films, i.e. to lipid-water interfaces. Some suggestive correlations may be obtained from the data of

TABLE 4

In vitro effects of tricyclic drugs on surface activity and membranes

Drug		Dynes/cm decrease in surface tension of saline by $10^{-3}M$ drug	Per cent haemolysis by $10^{-3}M$ drug	Per cent 5HT release from platelets by $3.10^{-4}M$ drug	Haemolysis 5HT release
Promazine	T	7	32	45	0.7
Chlorpromazine		12	60	49	1.2
Prochlorperazine		12	76	40	1.9
Perphenazine	↓	20	86	27	3.2
Opipramol		2	14	23	0.6
Amitryptyline	A	8	30	50	0.6
Imipramine		7	8	39	0.2
Nortryptyline		7	18	75	0.2
Desipramine	↓	7	1	63	0.02

Data from Ahtee (1966a, b). Drugs are arranged in order of increasing tranquillizing (T) and antidepressive effectiveness (A), as Ayd (1961) and Bickel (1966) respectively.

Ahtee. Thus in Table 4 the series of tranquillizers: promazine (1), chlorpromazine (2), prochlorperazine (3), perphenazine (4) is in order of increasing tranquillizing effectiveness (Ayd, 1961) and also in an order of successive single structural alterations (i.e. $2 = 1$ with Cl in the R_2 position; $3 = 2$ with a piperazine group incorporated in the R_1 chain; $4 = 3$ but hydroxymethylated at the end of R_1). It is apparent that in this series surface, haemolytic and tranquillizing activity increase in parallel, while there is a suggestion of decreased 5HT release from platelets as tranquillizing activity increases. Therefore, the ratio of haemolytic activity to 5HT-releasing activity increases as tranquillizing activity increases.

Fluphenazine, which can be considered a fifth member of the series (= 4 with CF_3 instead of Cl at R_2) has a greater haemolytic activity than 4 (Mao and Noval, 1966) and has greater therapeutic effectiveness. The observation that inactive phenothiazines such as chlorpromazine sulphoxide and those with a quaternary nitrogen have negligible surface and haemolytic activity (Ahtee, 1966 a, b) is also consistent with the above correlations.

The antidepressant tricyclics, however, exhibit sharply different properties. Demethylation of amitryptyline and imipramine to the therapeutically active nortryptyline and desipramine results in decreased haemolytic activity and increased 5HT release from platelets while surface activity is unchanged. Thus antidepressant activity is associated with low ratio of haemolytic activity to 5HT-releasing activity. Opipramol has both antidepressive and tranquillizing effects and has a ratio between those of these two groups.

These results suggest that while both tranquillizing and antidepressive drugs have a tendency to bind to components of low polarity, such as membrane lipids, their therapeutic effects may involve binding at different specific sites. The erythrocyte and platelet surfaces might be considered as models for these sites. There is evidence that 5HT receptors are of low polarity, being probably gangliosides (Woolley and Gommi, 1964, 1965), although nothing is known about the nature of other receptors.

PHYSICOCHEMICAL ASPECTS OF PHENOTHIAZINE ACTION

While the above properties of the tricyclic drugs play an important role it is clear that activity is not simply due to affinity for membranes and to the possession of a side chain with an amino group. If this were so then the therapeutic properties which these drugs possess would be far more widespread. Other factors must also be involved. It would seem reasonable that the striking therapeutic properties of phenothiazines have a relationship with their most striking physicochemical property, i.e. the almost unique state of electrons in phenothiazine molecules. This may be appreciated by considering the energy levels at which electrons may exist in a molecule. As energy is in quanta there is not a continuum of possible energies which these electrons may have but a series of fixed permitted levels. These may be roughly calculated for any molecule from electron energy levels in its constituent atoms. From these calculated levels the so-called K values may be approximately obtained (Pullman and Pullman, 1958). Any molecule has a series of positive and negative K values corresponding to the various levels of energy at which electrons may exist in it. The lowest positive value is associated with the highest energy level in the molecule. The lower it is, the easier it is for an electron to get out of the molecule (i.e. the lower the ionization potential). Conversely the negative values of K are normally associated with energy levels in the molecule which are unoccupied by an electron and the smallest negative K is an approximate measure of the ease with which the molecule can gain an electron (i.e. the electron affinity). Szent-Györgyi (1960) gives a comprehensive list of K values for molecules of biological interest. Phenothiazines are unusual in that the electrons with most energy in these molecules were calculated to have K values of less than zero, which was thought to indicate extremely easy loss of these electrons. Phenothiazines appeared to have the unique property of being the only known substances having negative K values in their ordinary stable states. In the calculations of the Pullmans and of Szent-Györgyi the effect of the sulphur atom on the state of the electrons (Longuet-Higgins, 1949) was, however, not taken into account. When this is allowed for, small positive K values of about +0.20 are obtained (Orloff and Fitts, 1961). These are intuitively more reasonable and do not affect the deduction that a phenothiazine molecule should be able to lose an electron with extreme ease. This has now been directly substantiated by a photoelectric method (Lyons and Mackie, 1963). In general, easy electron donation is equivalent to easy oxidizability or free radical formation. A related property is that if phenothiazines are mixed with substances able to accept an electron and with shapes such that the electron while it is in the donor molecule can closely approach the acceptor in a suitable orientation, then what is called a charge-transfer complex can occur (Mulliken, 1952). In a charge-transfer complex an electron from a donating molecule is shared with an acceptor molecule and forms a bond between them. The electron may then move right across to the acceptor no longer linking the two molecules. Phenothiazines have been shown to form these complexes with various acceptor molecules (Beukers and Szent-Györgyi, 1962; Foster and Hanson, 1966). The biochemical significance of charge-transfer is that it is a mechanism by which electrons may be transferred with hardly any energy being needed as there is hardly any change in the geometry of the molecules—what Szent-Györgyi (1960) has called 'biochemistry without chemistry'. It has been suggested that phenothiazines may cause tranquillization by affecting the charge on the electric double layer of cell surfaces by a charge-transfer process (Karreman, Isenberg and Szent-Györgyi, 1959). This might be involved in the effect of chlorpromazine in increasing resting membrane potentials in guinea-pig cerebral cortex and in opposing the decrease of potential due to electric pulses (Hillman, Campbell and McIlwain, 1963).

Another possibility is that therapeutic effects may be due to charge-transfer at receptor sites for neurohormones. It is of interest that the neurohormonally active substances DA, NA and 5HT are all good electron donors and this property may well be involved in their interaction with receptors. Structural modifications of these neurohormone molecules which decrease electron donating power generally decrease pharma-cological activity.

Thus, to summarize the above; the tricyclic drugs have a number of important general properties which, it can be assumed, are involved in determining their mechanisms of therapeutic action. These properties are:

(a) Non-polarity in areas of their molecules which allows interaction with and transport through lipid structures such as membranes and possibly also allows binding at receptors.

(b) The R_1 side chain which is similar to the C—C—N side chain of neurohormones may be necessary for interaction with neurohormone receptors, etc.

(c) High electron donating power leading to charge-transfer properties which may play a part in processes occurring subsequent to binding at membranes and receptors.

EFFECTS ON BRAIN AMINE METABOLISM

Brain amines such as the catecholamines and 5HT probably play an important part in the regulation of behaviour (see Kety 1966, 1968 for reviews). There has, therefore been much work on the effects of tricyclic drugs on these substances. Possible interactions are manifold and to assess their significance adequately it would be necessary to investigate many problems, e.g. the time courses of concentrations of drugs and their metabolites in various parts of the brain and their relationships with bio-chemical and behavioural effects, the relative biochemical effects of drugs with qualitative and quantitatively different therapeutic properties, effects on different metabolic pools and in different brain areas. Only a part of this data is at present available. Also the neurochemical significance of each amine is a subject of controversy although there is now a body of evidence indicating NA to be a central neurotransmitter (Glowinski and Baldessarini, 1966; Salmoiraghi, 1966).

The effects of tricyclic drugs on NA may be interpreted using a model put forward by Brodie and his co-workers (see, for example, Costa, Boullin, Hammer, Vogel and Brodie, 1966) which accords with the results of many kinds of experiment. According to this model, the adrenergic neurone contains NA within granules where it is synthesized and strongly bound. The granule NA is in equilibrium with NA in the cytoplasm which can leak out of the cell and become attached to receptors. This process would result in receptors being continuously occupied by NA and is opposed by an active mechanism for transport of NA back into the cell. Also NA may be metabolized within the cytoplasm by mono-amine oxidase to deaminated metabolites and extraneuronally by O-methyltransferase to methylated metabolites (Kopin, 1966). Thus the contributions of neuronal and extraneuronal metabolism to NA de-toxication may be assessed by the relative rates of formation of deaminated and methylated metabolites. Transport of NA back into the cell may be inhibited by nerve impulses, thus causing a net increase in NA release.

Changes of brain amine metabolism following phenothiazine adminis-tration have been known for a number of years. Thus, Gey and Pletscher (1961) found that 20 mg/kg chlorpromazine or chlorprothixene given to rats opposed the release of 5HT, NA and DA which occurs after reserpine administration. The increase of 5HT and NA due to monoamine oxidase inhibitors was also opposed as was the increase of 5HT due to injection of its precursor 5-hydroxytryptophan. The enzymes involved in amine metabolism are not affected by phenothiazines which suggested that the above effects were due to alterations of amine storage or transport. Somewhat similarly the stress-induced depletion of rat brain NA was prevented by 10 mg/kg chlorpromazine or phenobarbitone (Maynert and Levi, 1964). This is not simply a consequence of sedation, as it did not occur when strongly depressant doses of morphine were given. Caution is necessary, however, in the interpretation of the mechanism by which these high doses of phenothiazines act on amine metabolism, as Costa, Gessa and Brodie (1962) found that inhibition by chlorpromazine of amine release after reserpine was secondary to the hypothermic effect of chlorpromazine and did not occur when the rats were kept warm at 35°. Therefore, Carlsson and Lindqvist (1963) gave mice smaller doses of chlorpromazine (2.5–5 mg/kg) or haloperidol which is not a tricyclic compound but has therapeutic effects similar to those of chlorpromazine. These doses did not cause hypothermia, but resulted in increased brain levels of the methylated DA and NA metabolites, 3-methoxytyramine and normetanephrine. Promethazine, a phenothiazine with little tran-quillizing action, did not result in such changes. It was suggested that the tranquillizing drugs blocked amine receptors, which caused a compensatory activation of the monoaminergic neurones with increased release of monoamines resulting in increased subsequent formation of their extraneuronal metabolites. As the level of monoamines in the brain remains constant, an increased rate of synthesis is also implied. The findings that when catecholamine synthesis is blocked, chlorpromazine causes NA and DA depletion (Corrodi, Fuxe and Hokfëlt, 1967) and that nerve stimulation causes increased amine release and synthesis (Austin, Livett and Chubb, 1967) are consistent with this interpretation. The effects of antidepressive drugs were not investigated by Carlsson and Lindqvist. However, Glowinski and Axelrod (1964) found that when imipramine, desipramine or amitryptyline were given to rats, uptake of NA by brain was decreased. There is evidence that this is due to effects of the drugs on the granule membrane (Costa *et al*, 1966) and also on the nerve cell membrane (Carlsson, Fuxe, Hamberger and Lindqvist, 1966). Chlorpromazine and therapeutically inactive substances structurally related to imipramine were found by Glowinski and Axelrod to have no effect. This suggests an essential biochemical opposition between the antidepressants and the tranquillizing phenothiazines, the former decreasing NA uptake and thus making more available for action at receptors and the latter blocking receptors.

However, there are many reported qualitative similarities between effects of tricyclic depressants and tranquillizers. For example, in peripheral nervous tissue both imiprimine and chlorpromazine block NA uptake (Axelrod, Whitby and Hertting, 1961), chlorpromazine blocks uptake in brain slices (Dengler, Spiegel and Titus, 1961) and both chlorpromazine and imipramine inhibit the electrically stimulated NA release from rat brain slices (Baldessarini and Kopin, 1967). Also chlorpromazine, imipramine and desipramine all cause comparable inhibition of 5HT uptake by brain slices (Blackburn, French and Merrills, 1967).

There may be considerable differences in biochemical effects of the same drug on different parts of the brain. Thus Glowinski, Axelrod and Iversen (1966) found that if rats were given desipramine intraperitoneally before intraventricular administration of a small amount of ^{3}H-NA then there was decreased NA uptake in the cerebellum, medulla oblongata and hypothalamus but little effect on the striatum, mid-brain, hippocampus and cortex.

When ^{3}H-NA is already bound within the brain then both chlorpromazine and desipramine were found by Glowinski and Axelrod (1966) to inhibit its release. However, Schanberg, Schildkrant and Kopin (1967) found that while imipramine and desipramine inhibited release of ^{3}H-NA chlorpromazine did not. This difference cannot be easily interpreted without more kinetic information on neuronal incorporation and metabolism of ^{3}H-NA. Another disagreement with Glowinski and Axelrod's earlier work was that both antidepressants and chlorpromazine were found to inhibit brain NA uptake. Schanberg *et al* (1967) suggest this discrepancy may be due to the failure of the earlier workers to counteract hypothermia due to chlorpromazine. It is of particular interest that they found lithium chloride to decrease formation of ^{3}H-methylated metabolites but to increase formation of ^{3}H-deaminated metabolites. This suggests that the beneficial effort of lithium on mania may involve less NA being available extraneuronally for receptor binding. Reserpine caused similar changes in amounts of metabolites (Glowinski *et al*, 1966).

Parkinsonism and the biochemical pharmacology of phenothiazines
In Parkinson's disease a marked alteration of brain catecholamine metabolism occurs and there are indications that related biochemical changes may occur in phenothiazine-induced parkinsonism. Hornykiewicz and his co-workers (Barolin, Bernheimer and Hornykiewicz, 1964; Bernheimer and Hornykiewicz, 1965; Ehringer and Hornykiewicz, 1960; Hornykiewicz, 1963, 1966) found very low concentrations of NA and of DA, its precursor, in the basal ganglia of subjects with Parkinson's disease or post-encephalitic parkinsonism. The low DA values are particularly significant as this substance is sharply localized in the basal ganglia and substantia nigra, there being a dopaminergic neurone system with cell bodies in the substantia nigra having their terminals in the caudate nucleus and putamen (Andén, Dahlstrom, Fuxe and Larsson, 1965). Furthermore, administration of dopa, the precursor of both DA and NA, has been reported to alleviate parkinsonism, while reserpine, which depletes these substances, may cause parkinsonian symptoms. These considerations lead to the conclusion that DA is not merely a precursor of NA but is itself a neurohormone necessary for basal ganglia functions (see Curzon, 1968, for review).

It is reasonable to expect that the effects of drugs on DA and NA will be different as these substances are localized differently. Thus, for example, while desipramine reduced NA uptake, it did not reduce DA uptake (Glowinski *et al*, 1966; Carlsson *et al*, 1966). When DA metabolites were determined without ^{3}H-amine administration it was found that chlorpromazine and haloperidol elevated the DA metabolites dihydroxyphenylacetic and homovanillic acids in rabbit corpus striatum (Andén,

Roos and Werdinius, 1964), DA level being unchanged. Raised striatal homovanillic acid content occurs after administration of many tranquillizers, not all of which are phenothiazines (Roos, 1965). It is of interest that injection of oxotremorine, a non-phenothiazine, which causes parkinsonian-like tremor in animals results in increased homovanillic acid in the cat caudate nucleus (Laverty and Sharman, 1965). Also only phenothiazines which affect extrapyramidal function cause increased rat brain homovanillic acid (Da Prada and Pletscher, 1966).

There are two explanations of an increase in brain of homovanillic acid, the terminal dopamine metabolite. Either transport of homovanillic acid away from the brain is inhibited or both formation and breakdown of dopamine are increased by phenothiazines. Da Prada and Pletscher found that chlorpromazine did not cause elevation of the rise of homovanillic acid in brain which occurred after release of amines by a reserpine-like substance, hence, transport of homovanillic acid is unaffected. The suggestion of Carlsson and Lindqvist (1963) that the increase was a consequence of the blocking of dopamine receptors leads to a unitary hypothesis in which symptoms in Parkinson's disease are related to the lack of DA in the striatum while drug-induced parkinsonian symptoms are related to depletion of DA in the case of reserpine and to blockade of DA receptors in the cases of phenothiazines and possibly of oxotremorine.

Phenothiazines, as well as antagonizing the effects of catecholamines, may also after repeated dosage potentiate them (Webster, 1965). Furthermore, antagonism of central effects of 5HT and tryptamine (Tedeschi, Tedeschi and Fellows, 1961) has been reported. Antihistamine action may also occur. Hence, it would be unwise to put too much emphasis on any single explanation of the effects of these drugs.

Some phenothiazines such as diethazine which have R_1 chains with a C—C—N type structure have antiparkinsonian effects. This may relate to anti-acetylcholine properties. White and Westerbeke (1961) found that only phenothiazines which had antiparkinsonian action inhibited the effect of cholinergic drugs on the EEG. On the other hand, the anti-acetylcholine potency of a series of phenothiazines did not correlate with their antiparkinsonian effects (Ahmed and Marshall, 1962).

PHENOTHIAZINE MELANIZATION AND RELATED PHENOMENA

Melanization of the skin has been reported in patients receiving chlorpromazine for long periods (Greiner and Berry, 1964) and also in the viscera of a number of the above patients who died suddenly (Greiner and Nicholson, 1964). A number of different properties of phenothiazines could be invoked to explain this. Phenothiazines cause release of melanocyte-stimulating hormone from the rat pituitary (Kastin and Schally, 1966). Greiner and Nicholson observed pineal atrophy or necrosis in their patients which may be significant as melatonin, a melanocyte lightening factor, is synthesized in the pineal (Lerner, Case and Heinzelman, 1959). The pigment may also contain non-melanin material which may in part be derived from chlopromazine itself. Chlorpromazine accumulates in melanin-containing tissues, including skin (Forrest, Forrest and Roizin, 1963; Blois, 1965), probably due to a charge-transfer reaction with melanin (Forrest, Gutmann and Keyzer, 1966) which is a good electron acceptor (Commoner, Townsend and Pake, 1954). Satanove (1965) suggested that sunlight acting on phenothiazines in the skin was involved in the pigmentation as the latter increased through the summer months. Eventually a purple colour developed which did not resemble melanin. Recently, evidence has been reported that the metabolism of chlorpromazine in those subjects who respond by pigmentation is different from that in subjects who do not become pigmented. Thus, a major detoxication pathway of chlorpromazine in man is ring hydroxylation followed by conjugation (Bolt, Forrest and Serra, 1966). Patients with chlorpromazine pigmentation excrete abnormally large amounts of free hydroxylated chlorpromazine and abnormally low amounts of conjugates of hydroxylated chlorpromazine (Bolt and Forrest, 1967). Pigmentation in the eye, however, is not related to hydroxylated chlorpromazine formation.

Melanin formation may play a part in phenothiazine-induced extrapyramidal effects in view of the high concentration of melanin in the substantia nigra and the frequent finding of lesions in this area in Parkinson's disease (Greenfield and Bosanquet, 1953). Tyrosinase, the enzyme neces-sary for peripheral melanin formation, is a cuproprotein and it is known that the copper content of the brain rises in guinea-pigs after phenothiazine treatment (Holbrook, 1961). However, little is known about brain melanin formation and it is not clear whether brain melanin is identical to skin melanin or whether a copper enzyme is involved in its formation (see Curzon, 1968). The relationship between phenothiazine melanization and melanization in schizophrenia is also of interest and has been speculated on by Greiner and Nicholson (1965).

REFERENCES

AHMED A. and MARSHALL P.B. (1962) *Br. J. Pharmacol.* **18**, 247
AHTEE L. (1966a) *Annls Med. exp. Biol. Fenn.* **44**, 431

AHTEE L. (1966b) *Annls Med. exp. Biol. Fenn.* **44**, 453
ALBAUM H.G. and MILCH L.J. (1962) *Ann. N.Y. Acad. Sci.* **96**, 190
ANDÉN N.E., ROOS B.E. and WERDINIUS B. (1964) *Life Sciences* **3**, 149
ANDÉN N.E., DAHLSTRÖM A., FUXE K. and LARSSON K. (1965) *Am. J. Anat.* **116**, 329
AUSTIN L., LIVETT B.G. and CHUBB I.W. (1967) *Life Sciences* **6**, 97
AXELROD J., WHITBY L.G. and HERTTING G. (1961) *Science* **133**, 383
AYD F.G. (1961) *J. Am. med. Ass.* **175**, 1054
BALDESSARINI R.J. and KOPIN I.J. (1967) *J. Pharmacol* **156**, 31
BAROLIN G.S., BERNHEIMER H. and HORNYKIEWICZ O. (1964) *Schwiezer Arch. Neurol. Psychiat.* **94**, 241
BERNHEIMER H. and HORNYKIEWICZ O. (1965) *Klin. Wschr.* **43**, 711
BEUKERS R. and SZENT-GYÖRGYI A. (1962) *Recl Trav. Chim. Pays-Bas (Belg.)* **81**, 255
BICKEL M.H. (1966) In *Anti-depressant Drugs*, p. 3. Ed. GARATTINI S. and DUKES M.N.G. Excerpta Medica, Amsterdam and London
BICKEL M.H. and BRODIE B.B. (1964) *Int. J. Neuropharmacol* **3**, 611
BLACKBURN K.J., FRENCH P.C. and MERRILLS R.J. (1967) *Life Sciences* **6**, 1653
BLOIS M.S. (1965) *J. invest. Dermatol.* **45**, 475
BOLT A.G. and FORREST I.S. (1967) *Proc. west. Pharmacol. Soc.* **10**, 11
BOLT A.G., FORREST I.S. and SERRA M.S. (1966) *J. pharm. Sci.* **55**, 1205
BRUNE G.G., KOHL H.H., STEINER W.G. and HIMWICH H.E. (1963) *Biochem. Pharmacol.* **12**, 679
CARLSSON A. and LINDQVIST M. (1963) *Acta pharmac. tox.* **20**, 140
CARLSSON A., FUXE K., HAMBERGER B. and LINDQVIST M. (1966) *Acta physiol. scand.* **67**, 481
CHRISTENSEN J., FENG L., POLLEY E. and WASE A. (1958) *Federation Proceedings* **17**, 358
COMMONER B., TOWNSEND J. and PAKE G.E. (1954) *Nature (Lond.)* **174**, 689
CORRODI H., FUXE K. and HÖKFELT T. (1967) *Life Sciences* **6**, 767
COSTA E., GESSA G.L. and BRODIE B.B. (1962) *Life Sciences* **1**, 315
COSTA E., BOULLIN D.J., HAMMER W., VOGEL W. and BRODIE B.B. (1966) *Pharmacol. Rev.* **18**, 577
CURZON G. (1968) *Int. Rev. Neurobiol.* **10**, 323
DA PRADA M. and PLETSCHER A. (1966) *Experientia* **22**, 465
DENGLER H.G., SPIEGEL H.E. and TITUS E.O. (1961) *Nature (Lond.)* **191**, 816
EHRINGER H. and HORNYKIEWICZ O. (1960) *Klin. Wschr.* **38**, 1236
FORREST F.M., FORREST I.S. and ROIZIN L. (1963) *Agressologie* **4**, 259
FORREST I.S., GUTMANN F. and KEYZER H. (1966) *Agressologie* **7**, 147
FORREST I.S., SNOW H.L., ERICKSON G., GEITER C.W. and LAXSON G.O. (1966) *Proc. west. Pharmacol. Soc.* **9**, 18
FOSTER R. and HANSON P. (1966) *Biochim. Biophys. Acta* **112**, 482
GEY K.F. and PLETSCHER A. (1961) *J. Pharmacol.* **133**, 18
GLASKY A.J. (1963) *Federation Proceedings* **22**, 272
GILLETTE J.R., DINGELL J.V., SULSER F., KUNTZMAN R. and BRODIE B.B. (1961) *Experientia* **17**, 417
GLOWINSKI J. and AXELROD J. (1964) *Nature (Lond.)* **204**, 318
GLOWINSKI J. and AXELROD J. (1966) *Pharmacol. Rev.* **18**, 775
GLOWINSKI J. and BALDESSARINI R.J. (1966) *Pharmacol. Rev.* **18**, 1201

GLOWINSKI J., AXELROD J. and IVERSEN L.L. (1966) *J. Pharmacol.* **153**, 30

GORDON M., CRAIG P.N. and ZIRKLE C.L. (1964) In *Molecular Modification in Drug Design*, p. 140. Ed. GOULD R.F. American Chemical Society, Washington, D.C.

GREENFIELD J.G. and BOSANQUET F.D. (1953) *J. Neurol. Neurosurg. Psychiat.* **16**, 213

GREINER A.C. and BERRY K. (1964) *Canad. Med. Ass. J.* **90**, 663

GREINER A.C. and NICHOLSON G.A. (1964) *Canad. Med. Ass. J.* **91**, 627

GREINER A.C. and NICHOLSON G.A. (1965) *Lancet* **ii**, 1165

GREINER A.C., NICHOLSON G.A. and BAKER R.A. (1964) *Canad. Med. Ass. J.* **91**, 636

GUTH P.S. and SPIRTES M. (1964) *Int. Rev. Neurobiol.* **7**, 231

GUTH P.S., AMARO J., SELLINGER O.Z. and ELMER L. (1965) *Biochem. Pharmacol.* **14**, 769

HILLMAN H.H., CAMPBELL W.J. and McILWAIN H. (1963) *J. Neurochem.* **10**, 325

HOLBROOK J.R. (1961) *J. Neurochem.* **7**, 60

HORNYKIEWICZ O. (1963) *Wien. klin. Wschr.* **75**, 309

HORNYKIEWICZ O. (1966) *Pharmacol. Rev.* **18**, 925

HOSOYA S. (1963) *Acta Crystallogr.* **16**, 310

KARREMAN G., ISENBERG I. and SZENT-GYÖRGYI A. (1959) *Science* **130**, 1191

KASTIN A.J. and SCHALLY A.V. (1966) *Endocrinology* **79**, 1018

KETY S.S. (1966) *Pharmacol. Rev.* **18**, 787

KETY S.S. (1968) *Adv. Pharmacol.* In press

KOPIN I.J. (1966) *Pharmacol. Rev.* **18**, 513

KOPIN I.J. and GORDON E.K. (1962) *J. Pharmacol.* **138**, 351

LAVERTY R. and SHARMAN D.F. (1965) *Br. J. Pharmacol.* **24**, 759

LERNER A.B., CASE J.D. and HEINZELMAN R.V. (1959) *J. Am. Chem. Soc.* **81**, 6084

LONGUET-HIGGINS C. (1949) *Trans. Faraday Soc.* **45**, 173

LYONS L.E. and MACKIE J.C. (1963) *Nature (Lond.)* **197**, 589

MAO T.S.S. and NOVAL J.J. (1966) *Biochem. Pharmacol.* **15**, 501

MAYNERT E.W. and LEVI R. (1964) *J. Pharmacol.* **143**, 90

MULLIKEN R.S. (1952) *J. Am. Chem. Soc.* **74**, 811

ORLOFF M.K. and FITTS D.D. (1961) *Biochim. Biophys. Acta* **47**, 596

PHILLIPS B.M. and MIYA T.S. (1962) *Pharmacologist* **4**, 170

PULLMAN A. and PULLMAN B. (1958) *Proc. Nat. Acad. Sci.* **44**, 1197

RIBBENTROP A. and SCHAUMANN W. (1964) *Archs int. Pharmacodyn. Thér.* **149**, 374

ROOS B.E. (1965) *J. Pharm. Pharmac.* **17**, 820

SALMOIRAGHI G.L. (1966) *Pharmacol. Rev.* **18**, 717

SATANOVE A. (1965) *J. Am. med. Ass.* **191**, 263

SCHALES O. (1953) *Proc. Soc. exp. Biol. Med.* **83**, 593

SCHANBERG S.M., SCHILDKRAUT J.J. and KOPIN I.J. (1967) *Biochem. Pharmacol.* **16**, 393

SULSER F., WATTS J. and BRODIE B.B. (1961) *Ann. N.Y. Acad. Sci.* **96**, 279

SZENT-GYÖRGYI A. (1960) In *Introduction to a Submolecular Biology*. Academic Press, New York and London

TEDESCHI D.H., TEDESCHI R.E. and FELLOWS E.J. (1961) *Archs int. Pharmacodyn. Thér.* **132**, 172

VAN DEENEN L.L.M. and DEMEL R.A. (1964) *Biochim. Biophys. Acta* **94**, 314

WEBSTER R.A. (1965) *Br. J. Pharmacol.* **25**, 566

WHITE R.P. and WESTERBEKE E.J. (1961) *Expl. Neurol.* **4**, 317

WOOLLEY D.W. and GOMMI B.W. (1964) *Nature (Lond.)* **202,** 1074
WOOLLEY D.W. and GOMMI B.W. (1965) *Federation Proceedings* **24,** 367
ZOGRAFI G. and AUSLANDER D.E. (1965) *J. pharm. Sci.* **54,** 1313
ZÖLLER E., SCHREIER K. and YANG P.R. (1958) *Arzneimittel-Forsch.* **8,** 238

CHAPTER SEVEN

NEUROLOGICAL DISORDERS ASSOCIATED WITH ABNORMAL AMINO-ACID METABOLISM

O. H. WOLFF

I shall confine my remarks to the inherited disorders of amino-acid metabolism which involve the nervous system. Table 1 lists these disorders and their main symptoms. Singly these conditions are rare, but as a group they are not uncommon. Accurate diagnosis is crucial because for a few of them treatment is now available and for all of them the diagnosis carries important genetic implications.

To be successful treatment must be started early in infancy when clinical diagnosis is difficult or impossible and therefore mass screening programmes of newborn infants present a promising approach. The exact mechanisms whereby these disorders injure the brain are not understood; but it is clear that the immature brain of the infant and young child is unusually susceptible to all sorts of noxious agents.

Caution is needed before claiming a causal relationship between a biochemical abnormality, such as an abnormal aminoaciduria or elevated serum level of an amino acid, and a neurological abnormality such as mental retardation. Not all inborn errors of metabolism are harmful and chance determines that occasionally a harmless chemical abnormality will co-exist with, but not be causally related to, a neurological abnormality. In the case of some disorders listed in Table 1 only a very few patients have so far been reported and the clinical features listed in the third column may prove incomplete or even incorrect.

Table 2 lists the main neurological symptoms and signs, which, singly or jointly, should alert the clinician to the possibility of one of the inborn errors of amino-acid metabolism.

I shall give a brief account of the clinical features of three of the more common and treatable conditions: phenylketonuria, maple syrup urine disease (branched chain ketoaciduria) and homocystinuria.

TABLE I

Inherited disorders of amino-acid metabolism involving the nervous system

Amino acid	Disease	Clinical features
Glycine	Hyperglycinaemia	Ketosis Neutropenia Thrombocytopenia Hypo-gammaglobulinaemia Physical and mental retardation
Methionine	Methionine malabsorption	Mental retardation Convulsions Episodes of hyperventilation Intermittent diarrhoea
Cystathionine	Cystathioninuria	Mental retardation Congenital abnormalities Psychosis
Homocysteine	Homocystinuria	Mental retardation Dislocated lens Thrombo-embolic phenomena Convulsions Sparse hair
Valine	Hypervalinaemia	Mental retardation Vomiting Nystagmus
Valine Isoleucine Leucine	Maple syrup urine disease	Spasticity Mental retardation Convulsions
Leucine	Leucine-induced hypo-glycaemia	Hypoglycaemic convulsions
Phenylalanine	Phenylketonuria	Mental retardation Convulsions Eczema
Tryptophane	Congenital tryptophanuria with dwarfism	Mental and physical retardation Pellagra-like skin lesions Conjunctival telangiectasia Photosensitivity Ataxia
	Hartnup disease	Mental retardation Pellagra-like rash Cerebellar ataxia
	Hyperserotonaemia	Intermittent flushing Ataxia Convulsions

Tryptophane (*contd.*)	Hydroxykynureninuria	Mental and physical retardation Bloody diarrhoea Haemolytic anaemia

Histidine	Histidinaemia	(a) Normal
		(b) Speech disorders
		    Physical retardation
		    ? Recurrent infections

Arginine	Argininosuccinic aciduria	Mental retardation Convulsions Ataxia Vomiting Abnormal hair
Lysine	Hyperlysinaemia	Mental and physical retardation Hypotonia Fine hair Ammonia intoxication
Citrulline	Citrullinaemia	Mental retardation Vomiting Ammonia intoxication
Ornithine	Hyperammonaemia	Mental retardation Cortical atrophy Ammonia intoxication
Proline	Hyperprolinaemia Type I	 Renal disease Photogenic epilepsy Deafness Mental retardation
	Type II	Mental retardation Convulsions
Hydroxyproline	Hydroxyprolinaemia	Mental retardation Microscopic haematuria

TABLE 2

Symptoms and signs suggestive of an inborn error of amino-
acid metabolism

Mental retardation
Convulsions
Ataxia
Pellagra-like rash
Eczema
Abnormalities of hair (sparse, brittle)
Dislocated lenses
Physical retardation
Speech disorders
Attacks of vomiting and coma (ammonia intoxication)

PHENYLKETONURIA

Recent surveys based on mass screening programmes suggest an incidence of one case of phenylketonuria in approximately 10,000 births.

The most important symptom is the mental defect which in the untreated patient is usually severe. Exceptionally an individual with all the chemical features of the disorder may be of normal intelligence. I have under my care a healthy 7½-year-old boy of above-average intelligence (I.Q. 130). I was consulted about him because his brother had recently died of leukaemia and the parents wished to be reassured about the health of their remaining son. A routine phenistix test was strongly positive and serum phenylalanine levels determined on three occasions were 30 mg per 100 ml. Such individuals may represent the upper end of the distribution curve of the intelligence in phenylketonuria. Another explanation is that there may be an allelic gene which has chemical effects similar to those of classical phenylketonuria but does not lead to mental deficiency (Woolf et al, 1961). A third hypothesis is that modifying genes render the effects of the metabolic lesion less severe (Pratt, 1967).

Convulsions are frequent and in infancy usually take the form of infantile spasms. In later childhood petit mal or grand mal seizures occur. Even in patients without seizures the electroencephalogram is usually abnormal.

Behavioural disturbances are frequent, even in patients whose intelligence is normal or near normal and treatment with a low phenylalanine diet may improve the behaviour. On the other hand, long-continued dietetic treatment often leads to, or aggravates, problems of behaviour.

Other neurological abnormalities which are occasionally present are increased muscle tone and tendon reflexes, tremors, athetotic movements and restlessness. In the severely affected patient microcephaly and psychotic features may be present.

Defective pigmentation is usual and most of these children are blue eyed and fair haired. Eczema may be a feature.

Though recently some doubt has been cast on the value of treatment with a low-phenylalanine diet (Birch and Tizard, 1967; Bessman 1967), the results obtained by workers such as Bickel (1967), Clayton, Moncrieff and Roberts, (1967) and Berry, Sutherland, Umbarger and O'Grady, (1967) show that treatment when started during the early weeks of life is effective. Berry et al (1967) found the mean intelligence quotient of eight children in whom treatment was started before 12 weeks of age to be 108 (range 102–118); the mean I.Q. of ten normal siblings was 106 (range 88–

139), and that of eight of the parents was 109 (range 93–126); but the mean I.Q. of five affected siblings in whom treatment was started after 1 year of age was only 68 (range 55–83).

Yet even with early diagnosis, during the first 2 or 3 months, and prompt initiation of well-controlled dietetic treatment, it is unusual for the child to develop an intelligence considerably above average, and the mean intelligence quotient in most series is in the region of 90. Such results, however, are not to be despised, for these children can have a normal education, whereas most untreated patients are so severely affected that they are unsuitable even for education at a school for educationally subnormal children.

When diagnosis is delayed, results of treatment are less satisfactory, but a trial of treatment should probably be undertaken even when the diagnosis has not been made until the age of 1 or 2 years. Some of these children, though their intelligence may not improve, will become more manageable in their behaviour, and epilepsy, if present, may become controlled.

The recognition of the importance of early diagnosis led to the setting up of schemes for mass screening of all newborn infants in many parts of the world. This country was among the first to introduce such a programme, based on the phenistix test, which depends on the presence in urine of phenylpyruvic acid, one of the metabolites of phenylalanine. The introduction of this test was a real advance, but with greater experience it has become clear that the test has theoretical and practical disadvantages and fails to detect a considerable number of patients. Thus Clayton *et al* (1967) in a study of fifty-seven patients referred to the Hospital for Sick Children, Great Ormond Street, found that in only seven was the condition detected by routine testing. Most of the remaining patients were referred for the investigation of such conditions as mental subnormality (32), infantile spasms (4), eczema (2). This series also includes seven infants known to be 'at risk' because of a sibling in whom phenylketonuria had already been diagnosed. In another recent study (Stephenson and McBean, 1967b) the detection rate with the phenistix test was only 27 per cent. Fortunately other more sensitive screening tests have now been devised which give a positive result already at the age of 6 days provided of course the infant is receiving adequate amounts of milk, whereas the phenistix test may not become positive until the age of 3–4 weeks or occasionally even later. Of these tests, the Guthrie blood method (Guthrie and Susi, 1963) which depends on bacterial inhibition, has been used most widely and, under the conditions peculiar to a mass screening programme, has been found reliable, simple and cheap. Other promising methods which have not yet been tried out quite so extensively in mass trials include chromatography of amino acids in blood (Berry, 1962; Efron, Young, Moser and MacCready, 1964; Scriver, Davies and Cullen, 1964; Mellon and Stiven, 1966), spectrophotofluorimetry of blood (Hill, Summer, Pender and Roszel, 1965; Searle, Mijuskovic, Widelock and Davidow, 1967) and urine chromatography for o-hydroxyphenylacetic acid (Woolf, 1967b). As Woolf (1967a) points out, these various methods, except spectrophotofluorimetry, have the advantage that they can be adapted to detect other treatable inborn errors of metabolism such as maple syrup urine disease and histidinaemia.

The advent of these screening methods has brought up new problems, in particular that of the interpretation of the positive screening test. It has become clear that not every high serum level of phenylalanine is due to phenylketonuria and that a variety of hyperphenylalaninaemias exist (Scriver, 1967; Stephenson and McBean, 1967a). A positive screening test is therefore only the first step towards diagnosis and should lead to the immediate referral of the patient to a hospital fully equipped for the investigation of these metabolic disorders. The diagnosis and management of phenylketonuria requires a team approach, the team consisting of biochemist, dietitian, psychologist and paediatrician. Occasionally it is necessary in an infant who is already being treated with a low phenyl-alanine diet to reconfirm the diagnosis during the later months of the first year by giving a phenylalanine load and following the blood levels of the amino acid.

The details of treatment have been described by Clayton, Francis and Moncrieff (1965), Bickel and Bremer (1967), Berry *et al* (1967) and by others. It is not yet possible to state how high a serum phenylalanine level is permissible. Some workers aim at levels below 5 mg per 100 ml, others regard levels below 8 mg as satisfactory; levels below 1 mg are dangerous and are responsible for rashes, loss of appetite, wasting and even death (Moncrieff and Wilkinson, 1961).

Another question to which at present no clear answer can be given concerns the age at which dietary treatment should be stopped: the diet is unpleasant and as they grow older most children tire of it and rebel against it. Serious emotional problems occur occasionally. In children in whom treatment was started too late to achieve real benefit the decision to stop the diet after a trial period of 1–2 years is not difficult. When there has been a good response to the diet, some workers advise life-long dietary treatment, other stop the diet at the age of 5 years. In agreement with the views of Bickel (1967) we have discontinued the diet in a few children around the age of 10 years, and so far have not regretted this step. The introduction of the normal diet should be carried out under psychometric and electroencephalographic control (Clayton, Moncrieff, Pampliglione and Shepherd, 1966). In the patient who has epilepsy it may be necessary to continue the low-phenylalanine diet indefinitely in order to control the fits.

When a woman with phenylketonuria becomes pregnant, the low-phenylalanine diet will have to be restarted as early in the pregnancy as possible in order to prevent damage to the brain of the foetus from the high concentrations of phenylalanine which will cross the placenta from mother to foetus. Mabry, Denniston and Coldwell (1966) described the occurrence of mental retardation in the thirteen children of three phenylketonuric mothers. Two of the mothers were of normal intelligence and none of the children had phenylketonuria.

MAPLE SYRUP URINE DISEASE

As in phenylketonuria, inheritance is through an autosomal recessive gene; the enzymic block concerns the oxidative decarboxylation of the keto-acids of the three branched-chain amino acids, valine, leucine and iso-leucine, with the result that these amino acids and their keto-acids are present in abnormally high concentrations in blood and urine. The condition usually progresses quickly and affected infants may die undiagnosed during the early days of life. Difficulty in feeding, anorexia, vomiting and drowsiness are early symptoms. Spasms of increased muscle tone alternating with hypotonia and convulsions occur. The urine has the smell of maple syrup already during the first week of life. Hypoglycaemic attacks are frequent (Donnell, Lieberman, Shaw and Koch, 1967). Most untreated patients die in early infancy; but the severity of the disease varies greatly and a few patients may survive with brain damage. Morris, Lewis, Doolan and Harper (1961) described a child in whom the biochemical changes were present only intermittently, accompanied infections and were associated with attacks of ataxia, lethargy and convulsions. Despite these episodes mental and physical development was normal.

Westall (1967) reported the case of a 5-year-old child in whom treatment with a diet low in leucine, isoleucine and valine has been started on the sixth day of life and who had developed normally, both mentally and physically. However, during periods of inadequate dietary control and during intercurrent infections resulting in breakdown of body protein, blood levels of the branched-chain amino acids became too high and the patient became ataxic, started to vomit, became dehydrated and occasionally even lapsed into coma. Snyderman (1967) also reported the successful treatment of a child in whom diagnosis was started at the age of 10 days. In five other patients, however, in whom the diagnosis was delayed until the age of 17 days or later, normal mental development was not achieved. Timely clinical diagnosis is difficult unless the typical smell is noticed and mass screening schemes of blood or urine during the first week of life will have an important place in the early detection of this treatable condition.

HOMOCYSTINURIA

Among the inborn errors of amino-acid metabolism, homocystinuria is probably second only to phenylketonuria in frequency. Inheritance is through an autosomal recessive gene. The primary abnormality is a deficiency of the enzyme cystathionine synthetase; as a result, homo-cysteine and homocystine are not converted into cystathionine and homocystine overflows into the urine. In some patients methionine also accumulates in blood and overflows into the urine.

A wide spectrum of symptoms and signs exists and some patients are only slightly affected, whilst others are severely handicapped. This wide variation makes evaluation of the various forms of dietary treatment difficult.

Mental retardation may be mild or severe and some affected children have a normal intelligence (Schimke, McKusick, Huang and Pollack, 1965). In infancy convulsions are frequent. Most of the patients have fine fair hair and a malar flush. Dislocation of the lenses is a typical finding and cataracts also occur. The skeletal changes are genu valgum, pes cavus, pes planus, hyphoscoliosis, unduly prominent joints and long fingers with periarticular thickening of the interphalangeal joints.

Thrombo-embolic phenomena have been reported in several of the patients and pulmonary embolus is responsible for the sudden deaths which may occur, particularly after administration of anaesthetics (Komrower and Wilson, 1963). This tendency to thrombo-embolic accidents may result from abnormal stickiness of platelets, and both increased and decreased numbers of platelets have been reported (Nyhan, 1967). Platelet abnormalities may be present in some of the other inborn errors of amino-acid metabolism such as hyperglycinaemia and cystathioninuria, and I have under my care a patient with phenylketonuria and thrombocyto-penia. The mechanisms responsible for these platelet abnormalities are not understood.

Dietary treatment may help these children and Waisman (1967) and Gaull (1967) discuss the theoretical considerations underlying such an approach. Ideally, treatment should probably consist of a diet rich in cystine and cystathionine but containing only little methionine. Komrower (1967) used a diet rich in cystine and poor in methionine; treatment was started on the ninth day of life and at the age of 2 years 4 months the patient's physical and mental development was normal, the eyes were normal but she had slight knock-knee and occasional reddening and puffiness of the hands.

SUMMARY

The inborn errors of amino-acid metabolism which are, or may be, associated with neurological abnormalities have been listed with their main symptoms and signs. Three conditions (phenylketonuria, maple syrup urine disease and homocystinuria), for which dietary treatment is now available, have been discussed in some detail. Treatment, in order to be effective, must be started early, at an age when clinical diagnosis may be difficult or impossible and mass screening programmes in the neonatal period offer a promising approach. Confirmation of the diagnosis and management of these disorders are not easy and are best undertaken in centres specializing in these rare disorders.

REFERENCES

BERRY H.K. (1962) *Am. J. ment. Defic.* **66,** 555
BERRY H.K., SUTHERLAND B.S., UMBARGER B., O'GRADY D. (1967) *Am. J. Dis. Child.* **113,** 2
BESSMAN S.P. (1967) *Am. J. Dis. Child.* **113,** 27
BICKEL H. (1967) In *Proceedings of the Conference on Phenylketonuria and Allied Metabolic Diseases, Washington. April 1966.* Children's Bureau, U.S. Government Printing Office
BICKEL H. and BREMER H.J. (1967) *Dt. med. Wschr.* **92,** 700
BIRCH H.G. and TIZARD J. (1967) *Develop. Med. Child Neurol.* **9,** 9
CLAYTON B., FRANCIS D. and MONCRIEFF A. (1965) *Brit. med. J.* **1,** 54
CLAYTON B., MONCRIEFF A.A., PAMPIGLIONE G. and SHEPHERD J. (1966) *Archs. Dis. Childh.* **41,** 267
CLAYTON B., MONCRIEFF A. and ROBERTS C.E. (1967) *Brit. med. J.* **2,** 133
DONNELL G.N., LIEBERMAN E., SHAW K.V.F. and KOCH R. (1967) *Am. J. Dis. Child.* **113,** 60
EFRON M.L., YOUNG D., MOSER H.W. and MACCREADY R.A. (1964) *New Engl. J. Med.* **270,** 1378
GAULL G.E. (1967) *Am. J. Dis. Child.* **113,** 103

GUTHRIE R. and SUSI A. (1963) *Pediatrics* **32,** 338
HILL J.B., SUMMER G.K., PENDER M.W. and ROSZEL N.O. (1965) *Clin. Chem.* **11,** 541
KOMROWER G.M. (1967) *Am. J. Dis. Child.* **113,** 98
KOMROWER G.M. and WILSON V.K. (1963) *Proc. R. Soc. Med.* **56,** 26
MABRY C.C., DENNISTON J.C. and COLDWELL J.G. (1966) *New Engl. J. Med.* **275,** 1331
MELLON J.P. and STIVEN A.G. (1966) *J. med. Lab. Tech.* **23,** 204
MONCRIEFF A. and WILKINSON R.H. (1961) *Archs Dis. Childh.* **1,** 753
MORRIS M.D., LEWIS B.D., DOOLAN P.D. and HARPER H.A. (1961) *Pediatrics* **28,** 918
NYHAN W. (1967) *Am. J. Dis. Child.* **113,** 108
PRATT R.T.C. (1967) In *Genetics of Neurological Disorders.* Oxford University Press, London
SCHIMKE R.N., McKUSICK V.A., HUANG T. and POLLACK A.D. (1965) *J. Am. med. Ass.* **193,** 711
SCRIVER C.R. (1967) *Pediatrics* **39,** 764
SCRIVER C.R., DAVIES E. and CULLEN A.M. (1964) *Lancet* **ii,** 230
SEARLE B., MIJUSKOVIC M.B., WIDELOCK D. and DAVIDOW B. (1967) *Clin. Chem.* **13,** 621
SYNDERMAN S.E. (1967) *Am. J. Dis. Child* **113,** 68
STEPHENSON J.B.P. and McBEAN M.S. (1967a) *Brit. med. J.* **3,** 579
STEPHENSON J.B.P. and McBEAN M.S. (1967b) *Brit. med. J.* **3,** 582
WAISMAN H.A. (1967) *Am. J. Dis. Child.* **113,** 101
WESTALL R.G. (1967) *Am. J. Dis. Child.* **113,** 58
WOOLF L.I., OUNSTED C., LEES D., HUMPHREY M., CHESHIRE N.M. and STEED G.R. (1961) *Lancet* **ii,** 464
WOOLF L.I. (1967a) *Brit. med. J.* **3,** 862
WOOLF L.I. (1967b) In *Proceedings of the Washington Conference on Phenylketonuria, 1966.* Ed. SWAIMAN K.F. and ANDERSON J.A. Washington D.C.

CHAPTER EIGHT

BIOCHEMICAL ASPECTS OF NEUROLOGICAL DISORDERS ASSOCIATED WITH ABNORMAL AMINO-ACID METABOLISM

L. I. WOOLF

The number of defects of amino-acid metabolism known to be associated with neurological disorder is now approaching thirty. Eventually several hundred may be discovered. Only a few conditions that illustrate general principles are considered here in any detail. All are inborn errors of metabolism in the sense in which the term was used by Garrod (1908), i.e. all involve a genetically determined defect in an enzyme or enzyme-like carrier. All enzymes are proteins and, as such, are coded for by chains of nuclear DNA which it is convenient to term genes. A change in one of these genes—a mutation—will either prevent the production of the relevant polypeptide or cause the production of a protein lacking catalytic properties, i.e. an inactive variant of the enzyme. An example is phenylketonuria.

PHENYLKETONURIA

Basic biochemistry

Classically, phenylketonuria is considered to be caused by the absence or inactivity of the enzyme phenylalanine hydroxylase. This enzyme catalyses the oxidation of phenylalanine to tyrosine (Fig. 1) and is present in liver; tyrosine hydroxylase, present in nerve cells, and can also catalyse this reaction but only at a very slow rate. If phenylalanine hydroxylase is inactive for any reason, phenylalanine accumulates and can reach concentrations in the body fluids of up to one hundred times normal.

The concentration of phenylalanine in the blood is normal at birth but rises rapidly because phenylalanine, alone among the amino acids, has its normal metabolic pathway blocked. Classically, the concentration levels off at 60–100 mg per 100 ml at the age of 8 or 9 weeks, but there is great variation in the rate of rise. After a few months the concentration falls again, gradually, to reach a stable level between 20 and 40 mg per 100 ml after about 2 years. It is not known why the concentration of phenyl-alanine in the blood falls from its early peak; it is certain that, in classical phenylketonuria, this fall is not the result of slow development of phenyl-alanine hydroxylase—this enzyme remains absent or totally inactive throughout life. Nor is it understood why, in different affected infants, the rate of rise of concentration of phenylalanine in the blood should vary so markedly.

Phenylalanine hydroxylase is not the only enzyme which acts on phenyl-alanine, but normally all reactions other than conversion to tyrosine are very slow. Examples are acetylation to produce *N*-acetylphenylalanine and transamination to produce phenylpyruvic acid (Fig. 2). Like most enzymatic reactions, the rate is, within limits, proportional to the concentration of substrate—when the concentration of phenyl-alanine rises to many times the normal value, the velocity of a reaction such as transamination becomes appreciable and a considerable proportion of the body's phenylalanine pool is converted to phenylpyruvic acid.

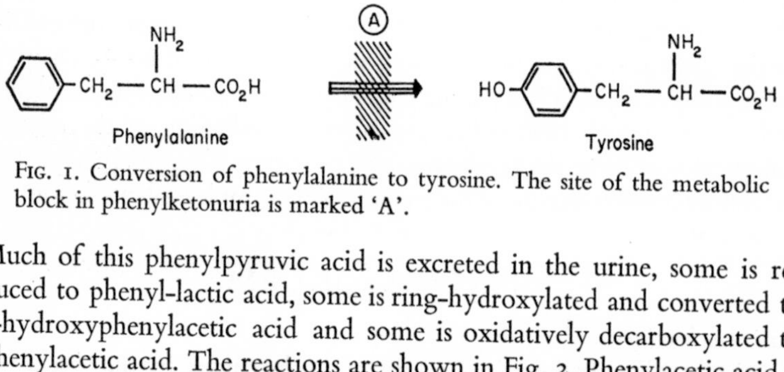

FIG. I. Conversion of phenylalanine to tyrosine. The site of the metabolic block in phenylketonuria is marked 'A'.

Much of this phenylpyruvic acid is excreted in the urine, some is re-duced to phenyl-lactic acid, some is ring-hydroxylated and converted to *o*-hydroxyphenylacetic acid and some is oxidatively decarboxylated to phenylacetic acid. The reactions are shown in Fig. 2. Phenylacetic acid is conjugated with glutamine and excreted in the urine as phenylacetyl-glutamine (Woolf, 1951); the other three acids are excreted uncon-jugated. The relative amounts of phenylalanine and its four major urinary metabolites excreted in phenylketonuria are shown in Table 1. This list of 'abnormal' metabolites of phenylalanine is not exhaustive, e.g. there is evidence that some phenyl-lactic acid is converted to benzoic acid and excreted as the glycine conjugate, hippuric acid (Armstrong, Chao, Parker and Wall, 1955); some phenylalanine is decarboxylated to phenethylamine (Jepson *et al*, 1960). These minor pathways account for only a small part of the total phenylalanine.

All these reactions are enzymic. It is not known whether the body possesses a specific phenylalanine transaminase or whether phenylalanine is acted on by a transaminase which, in the normal individual, acts on some other amino acid. It is no longer believed that enzymes are absolutely specific for a given substrate—phenylalanine and tyrosine are sufficiently

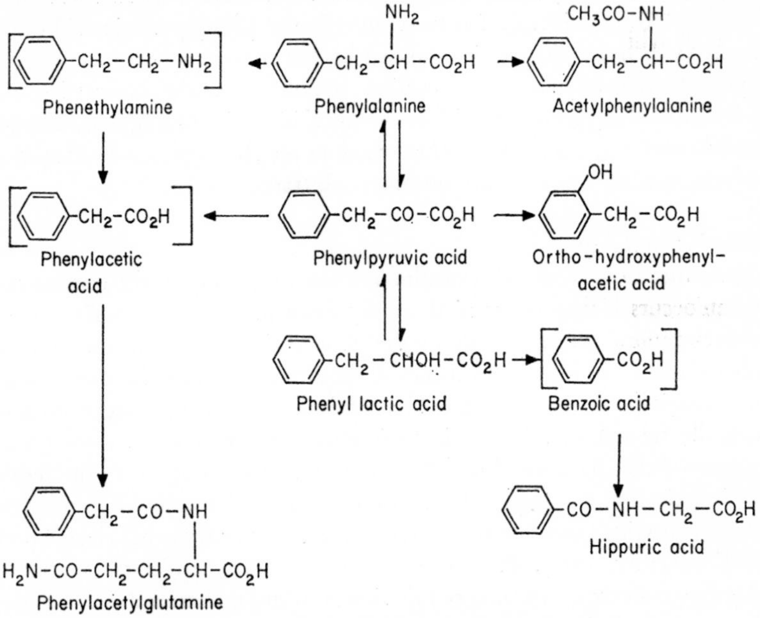

Fig. 2. Metabolism of phenylalanine in phenylketonuria. Major pathways are marked with heavy arrows, minor pathways with lighter arrows.

TABLE I

Urinary excretion of phenylalanine and its metabolites in μg/mg of creatinine. Range found in the majority of cases of classical phenylketonuria

Phenylalanine	300–1000
Phenylpyruvic acid	800–5600
Phenyl-lactic acid	600–3800
Phenylacetylglutamine	300–2400
o-hydroxyphenylacetic acid	100– 400

similar in structure (Fig. 1) to make it credible that tyrosine transaminase, for example, should act also on phenylalanine, slowly at normal concentrations, more rapidly when phenylalanine concentrations are elevated.

It is difficult to see what function a specific phenylalanine transaminase could serve in the mammalian economy. Similarly, the reduction of phenylpyruvic acid to phenyl-lactic acid (Fig. 2), known to be reversible, may be catalysed by lactic dehydrogenase rather than a specific enzyme. There is strong evidence that the conversion of phenylpyruvic acid to *o*-hydroxyphenylacetic acid is catalysed by the enzyme *p*-hydroxyphenyl-pyruvate hydroxylase in a reaction analogous to the conversion of *p*-hydroxyphenylpyruvic acid to homgentisic acid (Taniguchi, Kappe and Armstrong, 1964). The other reactions are also catalysed by enzymes which, normally, act mainly on other substrates.

Intoxication theory
These metabolites of phenylalanine are abnormal only in the concentrations in which they occur in the body fluids in phenylketonuria. These concentrations are the result of the abnormally high concentration of phenylalanine in blood and tissues. It was long held that the mental and neurological features and the biochemical abnormalities were coincidentally related, so-called pleiotropic effects of the gene. In 1951 it was suggested that the mental and neurological abnormalities were the result of intoxication by phenylalanine or its abnormal metabolites and that treatment with a diet low in phenylalanine might prove beneficial (Woolf and Vulliamy, 1951). The success, in many cases, of dietary treatment demonstrates the correctness of this view. It seems that most damage to the brain occurs during the first year of life when myelin is being laid down at the maximum rate, when the infant is learning fastest and most fundamentally about its environment and when the concentration of phenylalanine in the blood is at its highest. Very early restriction of phenylalanine intake seems to be essential if later intellectual development is to be normal. Later in childhood some phenylketonurics suffer gradual intellectual deterioration which is arrested or reversed by treatment with a low-phenylalanine diet. If dietary treatment is started later in childhood, there is often some improvement in behaviour, the EEG becomes more nearly normal and epileptic seizures, if present, may cease; these dramatic responses may occur within 3 or 4 days of starting treatment. However, it is rare for a late-treated phenylketonuric child to become intellectually normal.

Variants of phenylketonuria
For a long time the following simple picture was adequate: a mutant gene failed to produce phenylalanine hydroxylase and homozygotes for the

mutant gene could not convert phenylalanine to tyrosine; all were in consequence severely retarded and suffered other neurological disabilities. Then 'atypical phenylketonurics' of normal intelligence were discovered (Cowie, 1951; Coates, Norman and Woolf, 1957; Woolf, 1963). These atypical phenylketonurics usually, but not always, had normal EEGs and no history of epilepsy. In a few cases the I.Q. scores on verbal and performance scales were discordant (Woolf, Ounsted, Lee, Humphrey, Cheshire and Steed, 1961) and some came to clinical attention because of behavioural disorders (Sutherland, Berry and Shirkey, 1960). Some at least had about 7–10 mg of phenylalanine per 100 ml blood as against the typical phenylketonuric value of 20–40 (Woolf *et al*, 1961). Corresponding to this relatively low phenylalanine concentration, they excreted less phenylpyruvic acid, etc., than typical phenylketonurics. Recent work (Woolf, Goodwin, Wade and Patton, in preparation) indicates that atypical phenylketonurics convert phenylalanine to tyrosine at a rate considerably higher than typical phenylketonurics, though much lower than normal individuals or heterozygotes. It seems probable that, in atypical phenylketonuria, the relevant gene is coded for a variant enzyme with feeble, but measurable, catalytic properties; this has been confirmed by measurements on liver biopsy specimens (Justice, O'Flynn and Hsia, 1967). Normal intelligence probably implies that the brain was undamaged in infancy, i.e. that the concentration of phenylalanine in the blood and tissues was not much higher during the first year than in later life. Dietary treatment may be unnecessary in the unaffected, but may result in behavioural improvement (Sutherland *et al*, 1960) or increased learning ability (Coates, Norman and Woolf, 1957) in some. The late onset of epileptic seizures and mental deterioration in one atypical phenylketonuric is noteworthy (Woolf *et al*, 1961); it is possible that this could have been prevented by early diagnosis and treatment.

In some infants the concentration of phenylalanine in the blood rises during the first few days after birth to peak values of 10–50 mg per 100 ml (i.e. lower than in classical phenylketonuria) and then sinks to reach equilibrium levels of 3–10 mg per 100 ml by the age of 3–12 months (Scriver, 1967). This condition has been termed hyperphenylalaninaemia. There is both direct and indirect evidence that the liver contains some, relatively feeble, phenylalanine hydroxylase activity (Hudson, 1961; Stephenson and McBean, 1967; Justice *et al*, 1967; Woolf, Goodwin, Cranston, Wade, Woolf, Hudson and McBean, 1968). Recent evidence suggests that this phenylalanine hydroxylase is a variant form which loses activity rapidly at high concentrations of phenylalanine (Woolf *et al*,

1967). There are few clinical studies of untreated individuals: most seem to be moderately retarded (I.Q. 50 upwards) or normal intellectually; a few are severely retarded. It seems probable that damage to the brain occurs during the early phase of relatively high blood phenylalanine concentration and that dietary treatment is necessary only for the first few months in these cases.

Amyelination, dysmyelination and demyelination

Early histological examination indicated a lack of myelin in the white matter of the brain of some phenylketonurics but there was no general

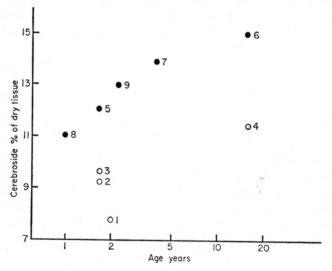

FIG. 3. Lack of myelin in phenylketonuria: cerebroside content of white matter plotted against age (log scale). ○, phenylketonurics; ●, normal controls. (Reprinted from *Journal of Neurology, Neurosurgery and Psychiatry*, 1962, Vol. 25, pp. 143–8, by permission of the editor and publishers.)

agreement as to whether this was a common feature (Alvord, Stevenson, Vogel and Engle, 1950; Poser and van Bogaert, 1959; Crome and Pare, 1960; Jervis, 1963). Chemical estimation of myelin lipids showed a deficiency in white matter from phenylketonuric brains as compared with age-matched controls (Fig. 3) (Crome, Tymms and Woolf, 1962; Menkes, 1966; Gerstl, Malamud, Eng and Kayman, 1967), though some authors failed to find this (Foote, Allen and Agranoff, 1965). In a few phenylketonuric brains there was a significant increase in cholesterol ester content of the white matter, this being confined to areas shown histologically to be undergoing active demyelination (Crome, 1962). It seems clear that phenylalanine or its abnormal metabolites can bring about destruction of myelin even during adult life, either by the destruction of neurones and consequent degeneration of the myelin sheaths around their axons or by damage to the oligodendroglia. In younger phenylketonurics there may be interference with the laying down of myelin rather than its destruction (though the latter cannot be ruled out) thus preventing the normal development of the brain during the first year or two of life.

Effects on synaptic transmission
The concentrations of adrenaline, noradrenaline and 5-hydroxytryptophan in the blood are reduced in phenylketonuria (Weil-Malherbe, 1955; Pare, Sandler and Stacey, 1957). It was found that DOPA-decarboxylase, necessary for the formation of noradrenaline and dopamine, 5-hydroxytryptophan decarboxylase, producing serotonin, and glutamic decarboxylase, producing gamma-aminobutyric acid, were inhibited by phenylalanine and its 'abnormal' metabolites (Fellman, 1956; Davison and Sandler, 1958; Hanson, 1959; Boylen and Quastel, 1961; Tashian, 1961). The products of these enzymes are neurohumoral transmitters in the central nervous system and the enzymes are concentrated at the synapses. It has been suggested that the mental and neurological features of phenylketonuria result from partial inhibition of these decarboxylases and, hence, reduced formation of transmitter substances (Fellman, 1956; Pare, Sandler and Stacey, 1958; Hanson, 1959; Tashian, 1961; Jervis, 1963).

Phenylalanine, in high concentration, inhibits the uptake of tyrosine and 5-hydroxytryptophan by brain cells (McKean, Schanberg and Giarman, 1962). These amino acids are precursors of dopamine, noradrenaline and serotonin—restriction of their entry into neurones could reduce the production of neurohormones. Such a reduction, however produced, would be rapidly reversible if the concentration of phenylalanine were to fall. This may account for the rapid improvement in behaviour, EEG and epilepsy in some older phenylketonurics as soon as they are treated with a low-phenylalanine diet, i.e. for the so-called reversible component of the disease process, the irreversible component being partly the result of structural damage to the brain during the first year of life (Woolf, Griffiths, Moncrieff, Coates and Dillistone, 1958; Crome, Tymms and Woolf, 1962).

Genetics
Phenylketonuria is a familial disease inherited as a Mendelian recessive

I

character (Penrose, 1935; Jervis, 1939; Munro, 1941, 1947). The heterozygotes are usually unaffected but tend to have a higher fasting plasma
phenylalanine level than normals: 1.13 ± 0.26 against 0.92 ± 0.28 mg per
100 ml. The overlap is too great for this to be a useful test for hetero-
zygosity, but the difference between the means is highly significant (Hsia
and Driscoll, 1956; Knox, 1963).

Giving a loading dose of phenylalanine, either orally (Hsia, Paine and
Driscoll, 1957; Jervis, 1960; Anderson, Gravem, Ertel and Fisch, 1962) or
intravenously (Woolf, Cranston and Goodwin, 1967) permits much better
discrimination. The intravenous method enables the quantity of active

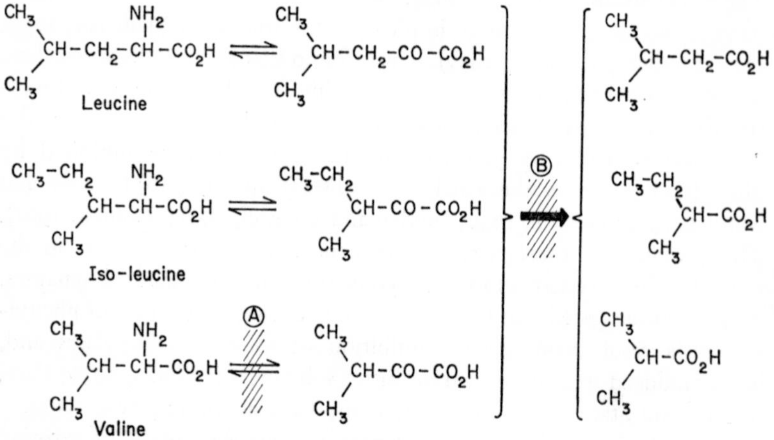

FIG. 4. Metabolism of leucine, isoleucine and valine. The metabolic block in
hypervalinaemia is marked 'A', that in maple syrup urine disease 'B',

enzyme per liver cell to be calculated; a heterozygote has, on the average,
half the normal amount of phenylalanine hydroxylase per liver cell. It is
possible, using the intravenous loading test, to classify correctly 94 per
cent of males and 99.2 per cent of females as heterozygotes or normals.

It seems probable that atypical phenylketonuria and hyperphenylalaninaemia are caused by different alleles at the phenylketonuria locus. The
gene coded for phenylalanine hydroxylase may have undergone several
different mutations, some leading to the production of inactive protein or
no protein at all in place of the enzyme, a second group coded for less
active forms of the enzyme and a third group coded for forms of the
enzyme unusually susceptible to inhibition or inactivation by high
concentrations of substrate. Homozygotes for the first group would have
classical phenylketonuria, those for the second group of mutations some

form of atypical phenylketonuria and those for the third group hyper-phenylalaninaemia (Zannoni *et al*, 1966; Justice *et al*, 1967; Woolf *et al*, 1967). Since all these mutations produce different alleles at the one locus, mixed genotypes and intermediate phenotypes are possible. For example, if a clinically unaffected individual with atypical phenylketonuria (or hyperphenylalaninaemia) were to marry a heterozygote for classical phenylketonuria, on average half the offspring would have phenyl-ketonuria of a biochemical type intermediate between classical and atypical phenylketonuria (or hyperphenylalaninaemia). It is very probable that such offspring would be severely mentally retarded if untreated. This situation may account for the reports of phenylketonuria in parent and child and of the apparent dominant inheritance of hyperphenylalaninaemia in some families.

MAPLE SYRUP URINE DISEASE (LEUCINOSIS) AND HYPERVALINAEMIA

The three branched-chain amino acids, leucine, isoleucine and valine are normally metabolized by very similar pathways (Fig. 4). Transmination to the three α-keto acids is catalysed by three separate enzymes, but the same enzyme system catalyses the oxidative decarboxylation of all three α-keto acids.

HYPERVALINAEMIA

A single case of hypervalinaemia has been reported in which valine transaminase was absent (Wada, 1965; Dancis, Levitz, Miller and Westall, 1967). The affected infant suffered from vomiting, failure to gain weight, mental retardation and hyperkinesis. The concentration of valine in the blood was 10.1 mg per 100 ml (normal value: 1.85) but there was no excess of α-keto-isocaproic acid. Absence of the transminase was proved by *in vitro* study of leucocytes. On a diet low in valine the child stopped vomiting, became less hyperkinetic and started to gain weight. However, a causal relationship between hypervalinaemia and the clinical features cannot be considered proven on a single case, nor do we know anything about the genetics of this condition.

MAPLE SYRUP URINE DISEASE (LEUCINOSIS)

Basic biochemistry

In leucinosis it was shown by Menkes (1959), and independently by two other groups (Mackenzie and Woolf, 1959; Davies, Levitz, Miller and

Westall, 1959), that there is a block in the oxidative decarboxylation of the three α-keto acids corresponding to leucine, isoleucine and valine (Fig. 4). This reaction is very complex, involving several enzymes and co-factors—it is not known which of these enzymes is defective in leucinosis. The three α-keto acids accumulate in the body fluids. Because transmination is a reversible reaction, the concentrations of leucine, isoleucine and valine in the body fluids rise to about ten times the normal. Part of each keto acid is reduced to the corresponding α-hydroxy acid which may undergo further chemical changes to produce the substances mainly responsible for the maple syrup odour.

Defects of myelination
The white matter of the brain is structurally grossly abnormal in leucinosis. Chemical analysis shows virtually complete absence of cerebrosides and sulphatides (Woolf, 1962; Menkes, 1967). It is as though formation of myelin were completely arrested within a few days of birth. This, of course, is consistent with the catastrophic onset of neurological signs in the neonatal period. In contrast, an affected infant treated with a diet low in the branched-chain amino acids had normal white matter when death occurred after intercurrent illness (Linneweh and Solcher, 1965).

Effect on brain enzymes
Tashain (1961) suggested that L-glutamic acid decarboxylase was in-hibited by the α-keto acids corresponding to valine, leucine and isoleucine and that the consequent deficiency of γ-aminobutyric acid affected synaptic transmission of impulses. Certainly, in leucinosis, the brain has only a quarter of the normal content of γ-aminobutyric acid (Prensky and Moser, 1966). However, Dreyfus and Prensky (1967) showed that glutamic acid decarboxylase was not affected by α-keto-isocaproic acid but that this acid, in the concentrations encountered in leucinosis, was a powerful in-hibitor of pyruvate decarboxylation. This indicates a fundamental effect on the Krebs cycle, the basic mechanism supplying energy to the cells of the central nervous system and elsewhere.

Genetics of leucinosis (maple syrup urine disease)
The disease is familial, occurring in siblings but never in parents and rarely in first cousins or more distant relatives. The pattern of occurrence is consistent with inheritance as a Mendelian recessive character. If this is so, the parents of an affected child should be heterozygotes for the relevant gene; leucocytes from the parents have reduced ability to decarboxylate

the three α-keto acids, as compared with normal controls. Leucocytes from patients with leucinosis lack this ability completely.

The prevalence of the disease is unknown, but it appears to be rarer than phenylketonuria. A figure of 1 in 200,000 live births has been suggested, but this seems too low in view of the low consanguinity rate among the parents of affected children; cases dying in the first few weeks of life may fail to be diagnosed.

Variant forms of the disease

Morris, Lewis, Doolan and Harper (1961) reported a patient with episodes of ataxia, lethargy and convulsions during which the α-keto acids corresponding to leucine, isoleucine and valine appeared in the urine, which smelt like maple syrup. Between attacks the urine was normal and the child never showed any retardation. Two siblings with a similar intermittent form of 'leucinosis' were described by Kiil and Rokkones (1964); in these children the neurological symptoms and excretion of branched-chain keto acids coincided with infections. The amount of decarboxylase present was presumably sufficient for ordinary metabolic needs and normally prevented the accumulation of excessive amounts of the three α-keto acids, etc., but, during infections, tissue breakdown would throw an overload of amino acids on the defective enzyme system, resulting in the accumulation of substrates and in neurological symptoms.

HISTIDINAEMIA

Basic biochemistry

Histidine is normally metabolized by conversion to urocanic acid and ammonia (Fig. 5), a reaction catalysed by the enzyme histidase which is widely distributed in the body. Urocanic acid is catabolized via formiminoglutaric acid and glutamic acid to α-ketoglutaric acid and so, via the Krebs cycle, to carbon dioxide and water. The enzyme histidase is absent in histidinaemia and histidine accumulates in the body fluids, reaching levels as high as 20 mg per 100 ml of blood (Ghadimi and Partington, 1967). The skin is normally rich in histidase and normal sweat contains urocanic acid; in typical histidiniaemia the skin lacks all histidase activity and the sweat is free from urocanic acid.

Transamination normally plays only a minor part in histidine metabolism, but the concentration of histidine in the tissues is so high that, in histidinaemia, much histidine is transaminated to imidazolepyruvic acid (Fig. 5). Part of this acid is excreted in the urine, part reduced to imidazolelactic acid and part decarboxylated to imidazoleacetic acid. There is a close parallel to phenylketonuria, even to the urine giving a green colour with phenistix or ferric chloride.

The way in which this metabolic error causes dysfunction of the central nervous system is unknown. There is no obvious correlation between the concentration of histidine in the blood and the degree of mental retardation in different cases, but more work is needed on the very wide range of

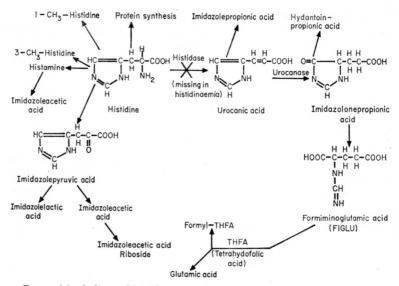

Fig. 5. Metabolism of histidine. The normal pathway to urocanic acid is blocked in histidinaemia and the alternative pathways, through imidazole-pyruvic acid, is therefore followed. (From *The Metabolic Basis of Inherited Disease, Histidinemia* by Bert N. La Du. Copyright 1960, 1966 by McGraw-Hill Inc. Used by permission of McGraw-Hill Book Company.)

I.Q. values in these children, the occurrence of specific defects, especially in speech, and the biochemistry.

Treatment with a diet low in histidine is being tried in a few cases, including a 2-month-old infant found in Oxford during neonatal screening for inborn errors of metabolism.[1] It is too soon to decide whether dietary treatment is effective. One difficulty, that does not arise with phenylketonuria, is that histidine appears not to be an essential amino acid for the adult human, i.e. we possess the ability to synthesize this amino

[1] The histidine-free food was generously provided by Trufood Ltd.

acid. Infants cannot synthesize sufficient histidine for their growth require-
ments and it must be supplied in the diet; we do not know when the
synthetic ability develops. Dietary restriction of histidine may be ineffective if the patient synthesizes the amino acid fast enough. Experience
in Oxford indicates that this is unlikely to occur before the age of 8
months, if at all.

Genetics
The occurrence of histidinaemia in siblings, but not in parents, suggested
inheritance as a Mendelian recessive character. Histidine loading tests, and
direct assay of skin histidase, in the parents produced further evidence that
they were heterozygotes for a mutant gene at the histidase locus (La Du,
1967).

INBORN ERRORS IN THE METABOLISM OF NON-ESSENTIAL AMINO ACIDS

There are nine essential amino acids that the human body requires to be
supplied in the diet, a further twelve occur in body proteins but are
endogenous in origin and yet others are biosynthesized and serve as
metabolic intermediates. Each of these non-essential amino acids is
normally catabolized by a series of reactions catalysed by specific enzymes.
Lack of one enzyme will cause its substrate to accumulate. An example is
hyperprolinaemia type II, where lack of the enzyme Δ'-pyrolline-5-
carboxylate dehydrogenase causes accumulation of proline accompanied
by convulsions and mental retardation (Efron, 1967). Dietary restriction
of proline would be useless since the body can synthesize proline so
rapidly.

There is, however, a group of such disorders in which treatment is
possible. Ammonia is converted to urea by the Krebs-Henseleit cycle (Fig.
6) involving the enzymes carbamyl phosphate synthase (1), ornithine
transcarbamylase (2), arginosuccinate synthase (3), arginosuccinase (4)
and arginase (5). Three inborn errors of metabolism involving enzymes
(2), (3) and (4) are known; they are, respectively, hyperammonaemia,
citrullinaemia and arginosuccinic aciduria (Levin and Russell, 1967;
Efron, 1966). All three are characterized by convulsions and mental
retardation and by a high level of ammonia in the blood. The clinical
features of all three conditions may result from ammonia intoxication.
Some heterozygotes suffer from migraine attacks and intermittent
hyperammonaemia. Treatment with a low-protein diet reduces the

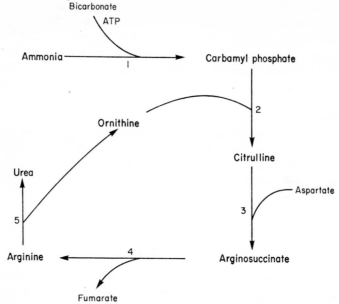

Fig. 6. Conversion of ammonia to urea by the ornithine cycle. Reaction 1 is catalysed by carbamoylphosphate synthase, 2 by ornithine carbamoyltransferase, 3 by arginiosuccinate synthetase, 4 by argininosuccinase and 5 by arginase.

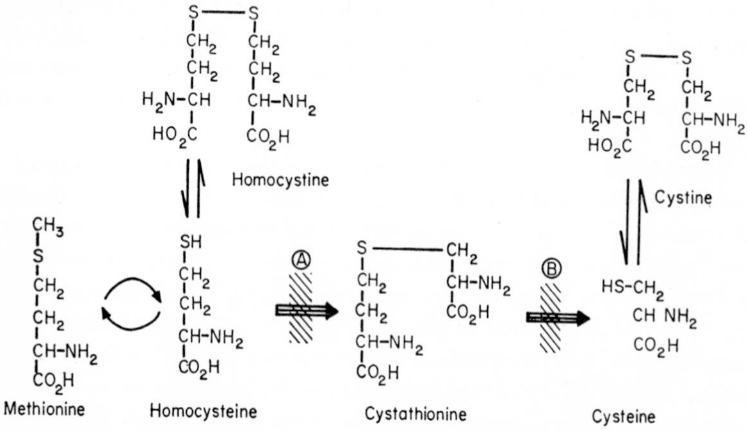

Fig. 7. Methionine metabolism. Demethylation of methionine to homocysteine is irreversible, but homocysteine is methylated by another system, hence the two amino acids are in equilibrium. The metabolic block in homocystinuria is marked 'A' and the metabolic block in cystathioninuria is marked 'B'.

concentration of ammonia in the blood and so does administration of citrate, aspartate or glutamate; such treatment leads to clinical improvement in some cases.

HOMOCYSTINURIA

Homocystinuria is, after phenylketonuria, the commonest inborn error of amino-acid intermediary metabolism. Methionine is an essential amino acid which is normally metabolized as shown in Fig. 7. As a rare inborn error of metabolism, some individuals lack cystathionine synthase and homocysteine accumulates. Methionine also often accumulates, reaching concentrations in the blood as high as 25–30 mg per 100 ml.

The absence of cystathionine synthase constitutes a block preventing the synthesis of products, such as cystationine and cystine, which are distal to the block. Much of the body's cystine is normally derived from methionine and there is a very real possibility that patients with homocystinuria may suffer from a lack of cystine. Cystathionine normally occurs in large amounts in nervous tissue, particularly the brain; it is virtually completely absent from the brains of homocystinurics (Gerritsen and Waisman 1964; Brenton, Cusworth and Gaull, 1965). It would not be easy to raise the concentration of cystathionine in the brain; oral or intravenous administration would lead to urinary excretion and catabolic breakdown rather than accumulation.

Nothing is known of the mechanism by which the characteristic lesions of homocystinuria come about. No biochemical difference was found, by Schimke, McKusick, Huang and Pollack (1965), between twenty-two mentally retarded homocystinurics and sixteen of normal intelligence. Any theory would have to account for this and for the rapid neurological deterioration suffered by some patients who may be mentally and neurologically normal up to the age of 2 or 3 years but complete aments suffering almost continuous epileptic seizures by 4 or 5 years of age. If, in the mentally unaffected, the level of cystathionine in the brain is as low as in those so far examined, this would indicate that cystathionine does not play an essential role in nerve cell metabolism. However, it is possible that while the enzyme cystathionine synthase is absent from the liver in all cases of homocystinuria, in some cases it may be present in other tissues such as the brain; it has been found in the lens of the eye in one case (Gaull, 1967).

On the theory that the clinical effects result from a toxic excess of methionine and homocysteine, treatment with a diet low in methionine has been tried (Brenton, Cusworth, Dent and Jones, 1966; Perry, 1966; Komrower, 1967); it is too soon to assess the results.

HARTNUP DISEASE

This represents a completely different type of inborn error of metabolism: instead of a defective enzyme involved in catabolism or anabolism, the mutant gene brings about a defect in a carrier system transporting certain amino acids into cells. It is believed that the amino acids in the renal glomerular filtrate are reabsorbed by four specific carriers in the proximal renal tubular epithelium, one for the imino acids proline, hydroxyproline (and glycine), one for the basic amino acids lysine, arginine, ornithine (and cystine), one for the acidic amino acids glutamic acid and aspartic acid, and one for a group of thirteen neutral amino acids. In Hartnup disease these neutral amino acids are reabsorbed inefficiently and the urine contains many times the normal amount of each, but only the normal traces of amino acids in the other three groups (Evered, 1956; Scriver, 1965; Jepson, 1966).

Like the renal proximal tubular epithelium, the jejunal mucosa is specialized for the absorption of various substances from the luminal contents. There is evidence that, as far as absorption of amino acids is concerned, the same four transport systems exist in the jejunal mucosa and the renal tubular epithelium. In Hartnup disease the thirteen amino acids excreted in excess in the urine are also poorly absorbed from the gut (Scriver, 1965); this was first shown specifically for tryptophan (Milne, Crawford, Girao and Loughridge, 1960). The poor absorption from the small intestine leads to unusually large amounts of tryptophan entering the colon where the bacterial flora produces indole, tryptamine, indole-acetic acid and other products which are absorbed (Shaw, Redlich, Wright and Jepson, 1960). Indole is oxidized in the liver to indoxyl which is excreted as the sulphate, indican; tryptamine is oxidized to indoleacetic acid; indoleactic acid, both bacterial and that derived from oxidation of tryptamine, is excreted partly free and partly conjugated with glutamine or glucuronic acid. Excessive amounts of these substances are present in urine from patients with Hartnup disease; the amounts rise during periods of clinical exacerbation and fall during periods of remission.

Tryptophan in the body is normally metabolized by several pathways (Jepson 1966); approximately 1.6 per cent is converted to nicotinamide, far outweighing in importance normal dietary sources of this vitamin. Nicotinamide forms part of the two co-enzymes, NAD and NADP, which play fundamental roles in many of the body's oxidation reduction processes. In Hartnup disease there is failure to absorb normal amounts of tryptophan coupled with excessive urinary loss of this amino acid, both factors tending to deprive the body of its source of nicotinamide. More important than either of these is the absorption of indole from the colon; Hooft, de Laey, Timmermans and Snoeck (1964) reported that indole was a potent inhibitor gf tryptophan-pyrollase and kynurenine-formamidase, two enzymes necessary for the conversion of tryptophan to nicotinamide. These workers found that production of nicotinamide and intermediates in Hartnup disease was increased by intravenous tryptophan but was actually decreased by orally administered tryptophan.

The clinical features of Hartnup disease are those of classical pellagra. Although sufferers from pellagra often have multiple dietary deficiencies, lack of nicotinamide is usually considered primary. It is reasonable to suppose that the signs and symptoms in Hartnup disease are also caused by nicotinamide deficiency and should be treated with this vitamin. Because high urinary excretion of indoles coincides with clinical exacerbation, and in view of the work of Hooft, de Laey, Timmermans and Snoeck (1964), oral neomycin or similar antibacterial agent should be given at such times (Woolf, 1966). Amine oxidase inhibitors may prevent oxidation of the toxic substance tryptamine to the far less toxic indoleacetic acid and should therefore be avoided in Hartnup disease (Milne, Asatoor and Loughridge, 1961).

CONCLUSIONS

The handful of diseases considered here permits some general principles to be put forward. In every case the mutation of a gene leads to a metabolic block, homozygotes for the mutant gene lacking the necessary enzyme or carrier substance or, in some cases, possessing altered enzymes of greatly reduced activity. At present no treatment is known by which a gene or its product, an enzyme or transport system, can be restored to normal, but recent advances suggest that within a few decades it may be possible to treat inborn errors of metabolism by organ transplants, so supplying the missing enzymes.

Absence of the enzyme or carrier system causes accumulation to toxic levels of the normal substrate or of its breakdown products; in some cases there may be a deficiency of a normal product of the blocked reaction. Where the substrate of the missing enzyme is not endogeneous, a rational and effective form of treatment is possible by restricting dietary content of the substrate or its precursor and by supplying missing reaction products. Many inborn errors of metabolism cause intellectual retardation. This is to be expected since the newborn infant, deprived of placental transfer and maternal metabolism, starts at once to accumulate toxic substances that interfere with the functioning of the brain at the time of its most rapid development. Damage to the brain in early infancy, whether or not it is demonstrable morphologically, invariably leads to mental retardation. We must not, however, lose sight of the dementia, psychotic behaviour and late onset of epileptic seizures seen in some of these conditions, evidence of continuing intoxication. The details of how the brain is attacked are mostly unknown, though they can be guessed at in some cases. This is the greatest single gap in our knowledge of these conditions.

REFERENCES

ALVORD E.C. Jr., STEVENSON L.D., VOGEL F.S. and ENGLE R.L. (1950) *J. Neuropath. exp. Neurol.* **9,** 298

ANDERSON J.A., GRAVEM H., ERTEL R. and FISCH R. (1962) *J. Pediat.* **61,** 603

ARMSTRONG M.D., CHAO F.-C., PARKER V.J. and WALL P.E. (1955) *Proc. Soc. exp. Biol. Med.* **90,** 675

BOYLEN J.B. and QUASTEL J.H. (1961) *Biochem. J.* **80,** 644

BRENTON D.P., CUSWORTH D.C., DENT C.E. and JONES E.E. (1966) *Q. Jl. Med.* **35,** 325

BRENTON D.P., CUSWORTH D.C. and GAULL G.E. (1965) *J. Pediat.* **67,** 58

COATES S., NORMAN A.P. and WOOLF L.I. (1967) *Archs Dis. Childh.* **32,** 313

COWIE V.A. (1951) *Lancet* **i,** 272

CROME L. (1962) *J. Neurol. Neurosurg. Psychiat.* **25,** 149

CROME L. and PARE C.M.B. (1960) *J. ment. Sci.* **106,** 862

CROME L., TYMMS V. and WOOLF L.I. (1962) *J. Neurol. Neurosurg. Psychiat.* **25,** 143

DANCIS J., HUTZLER J., TADA K., WADA Y., MORIKAWA T. and ARAKAWA T. (1967) *Pediatrics* **39,** 813

DANCIS J., LEVITZ M., MILLER S. and WESTALL R.G. (1959) *Brit. med. J.* **1,** 91

DAVISON A.N. and SANDLER M. (1958) *Nature (Lond.)* **181,** 186

DREYFUS P.M. and PRENSKY A.L. (1967) *Nature (Lond.)* **214,** 276

EFRON M.L. (1966) In *The Metabolic Basis of Inherited Disease,* Chap. 19. Ed. STANBURY J.B., WYNGAARDEN J.B. and FREDRICKSON D.S. McGraw-Hill Book Company, New York

EFRON M.L. (1967) *Am. J. Dis. Child.* **113,** 166

EVERED D.F. (1956) *Biochem. J.* **62,** 416

FELLMAN J.H. (1956) *Proc. Soc. exp. Biol. Med.* **93,** 413

FOOTE J.L., ALLEN R.J. and AGRANOFF B.W. (1956) *J. Lipid Res.* **6,** 518

GARROD A.E. (1908) *Lancet* **ii,** 1

GAULL G.E. (1967) *Am. J. Dis. Child.* **113,** 103

GERRITSEN T. and WAISMAN H.A. (1964) In *Proceedings of the International Copenhagen Congress on the Scientific Study of Mental Retardation*
GERSTL B., MALAMUD N., ENG L.F. and KAYMAN R.B. (1967) *Neurology (Minneap.)* **17,** 51
GHADIMI H. and PARTINGTON M.W. (1967) *Am. J. Dis. Child.* **113,** 83
HANSON A. (1959) *Acta chem. scand.* **13,** 1366
HOOFT C., DE LAEY P., TIMMERMANS J. and SNOECK J. (1964) In *Abstracts, First Meeting of the Federation of European Biochemical Societies, London*, p. 91
HSIA D.Y.-Y. and DRISCOLL K.W. (1956) *Lancet* **ii,** 1337
HSIA D.Y.-Y., PAINE R.S. and DRISCOLL K.W. (1967) *J. ment. Defic. Res.* **1,** 53
HUDSON F.P. (1961) *Brit. med. J.* **1,** 1105
JEPSON J.B. (1966) In *The Metabolic Basis of Inherited Disease*, Chap. 57. Ed. STANBURY J.B., WYNGAARDEN J.B. and FREDRICKSON D.S. McGraw-Hill Book Company, New York
JEPSON J.B., LOVENBERG W., ZALTMAN P., OATES J.A., SJOERDSMA A. and UDENFRIEND S. (1960) *Biochem. J.* **74,** 5
JERVIS G.A. (1939) *J. ment. Sci.* **85,** 719
JERVIS G.A. (1960) *Clin. Chim. Acta* **5,** 471
JERVIS G.A. (1963) In *Phenylketonuria*, p. 96. Ed. LYMAN F.L. Charles C. Thomas, Springfield
JUSTICE P., O'FLYNN M.E. and HSIA D.Y.-Y. (1967) *Lancet* **i,** 928
KIIL R. and ROKKONES T. (1964) *Acta Pediat.* **53,** 356
KNOX W.E. (1963) In *Phenylketonuria*, p. 11 Ed. LYMAN F.L. Charles C. Thomas, Springfield
KOMROWER G.M. (1967) *Am. J. Dis. Child* **113,** 98
LA DU B.N. (1967) *Am. J. Dis. Child* **113,** 88
LEVIN B. and RUSSELL A. (1967) *Am. J. Dis. Child* **113,** 142
LINNEWEH F. and SOLCHER H. (1965) *Klin. Wschr.* **43,** 926
MACKENZIE D.Y. and WOOLF L.I. (1959) *Brit. med. J.* **1,** 90
McKEAN C.M., SCHANBERG S.M. and GIARMAN N.J. (1962) *Science* **137,** 604
MENKES J.H. (1959) *Pediatrics* **23,** 348
MENKES J.H. (1966) *Pediatrics* **37,** 967
MENKES J.H. (1967) *Pediatrics* **39,** 297
MILNE M.D., ASATOOR A. and LOUGHRIDGE L.W. (1962) *Lancet* **ii,** 51
MILNE M.D., CRAWFORD M.A., GIRAO C.B. and LOUGHRIDGE L.W. (1960) *Q. Jl Med.* **29,** 407
MORRIS M.D., LEWIS B.D., DOOLAN P.D. and HARPER H.A. (1961) *Pediatrics* **28,** 918
MUNRO T.A. (1941) In *Proceedings of the International Genetics Congress (7th), Edinburgh*
MUNRO T.A. (1947) *Ann. Eugen.* **14,** 60
PARE C.M.B., SANDLER M. and STACEY R.S. (1957) *Lancet* **i,** 551
PARE C.M.B., SANDLER M. and STACEY R.S. (1958) *Lancet* **ii,** 1099
PENROSE L.S. (1935) *Lancet*, **ii,** 192
PERRY T.L., DUNN H.G., HANSEN S., MacDOUGALL L. and WARRINGTON P.D. (1966) *Pediatrics* **37,** 502
POSER C.M. and van BOGAERT L. (1959) *Brain* **82,** 1
PRENSKY A.L. and MOSER H.W. (1966) *J. Neurochem.* **13,** 863
SCHIMKE R.N., McKUSICK V.A., HUANG T. and POLLACK A.D. (1965) *J. Am. med. Ass.* **193,** 711

SCRIVER C.R. (1965) *New Engl. J. Med.* **273,** 530

SCRIVER C.R. (1967) *Pediatrics* **39,** 764

SHAW K.N.F., REDLICH D., WRIGHT S.W. and JEPSON J.B. (1960) *Federation Proceedings* **19,** 194

STEPHENSON J.B.P. and McBEAN M.S. (1967) *Brit. med. J.* **2,** 579

SUTHERLAND B.S., BERRY H.K. and SHIRKEY H.C. (1960) *J. Pediat.* **57,** 521

TANIGUCHI K., KAPPE T. and ARMSTRONG M.D. (1964) *J. biol. Chem.* **239,** 3389

TASHIAN R.E. (1961) *Metabolism* **10,** 393

WADA Y. (1965) *Tohoku J. exp. Med.* **87,** 322

WEIL-MALHERBE H. (1965) In *Proceedings of the 1st International Neurochemistry Symposium,* p. 458. Oxford

WOOLF L.I. (1951) *Biochem. J.* **49,** ix

WOOLF L.I. (1962) *Proc. R. Soc. Med.* **55,** 824

WOOLF L.I. (1963) *Adv. clin. Chem.* **6,** 97

WOOLF L.I. (1966) In *Renal Tubular Dysfunction.* Charles C. Thomas, Springfield

WOOLF L.I., CRANSTON W.I. and GOODWIN B.L. (1967) *Nature (Lond.)* **213,** 882

WOOLF L.I., GOODWIN B.L., CRANSTON W.I., WADE D.N., WOOLF F., HUDSON F.P. and McBEAN M.S. (1968) *Lancet* **i,** 114

WOOLF L.I., GOODWIN B.L., WADE D.N. and PATTON V.M. (1968) Unpublished

WOOLF L.I., GRIFFITHS R., MONCRIEFF A., COATES S. and DILLISTONE F. (1958) *Archs. Dis. Childh.* **33,** 31

WOOLF L.I., OUNSTED C., LEE D., HUMPHREY M., CHESHIRE N.M. and STEED G.R. (1961) *Lancet* **ii,** 464

WOOLF L.I. and VULLIAMY D.G. (1951) *Archs Dis. Childh.* **26,** 487

ZANNONI V.G., WEBER W.W., VAN VALEN P., RUBIN A., BERNSTEIN R. and LA DU B.N. (1966) *Genetics* **54,** 1391

CHAPTER NINE

HYPERTENSION AND CEREBROVASCULAR DISEASE

JOHN MARSHALL

Until just over 10 years ago cerebrovascular disease was a completely neglected field. At that time the availability of new diagnostic methods, such as angiography, and the development of potential therapy, such as anticoagulant and hypotensive drugs and endarterectomy, brought about what amounted to a revolution in this field. From a position of complete neglect, cerebrovascular disease moved to one of intense activity at the basic science, epidemiological and clinical levels. The initial gains were striking but, as in so many other fields of endeavour, early promise has not been entirely fulfilled. The first flush of enthusiasm and the rapid gains in knowledge have passed and we are now settling down to a period of unremitting toil in which further advance will be hard won. It seems appropriate, therefore, at this time to pause a moment to review the present status in the field of cerebrovascular disease and to see in which direction further advance is likely to be made. As the field is so great, it is necessary, even in a review, to limit the area of discussion, hence I propose to consider only the question of the relationship between hypertension and cerebrovascular disease.

PATHOLOGICAL CONCEPTS

In order that there may be no confusion about the entities we are discussing may I first remind you of some classical pathological concepts which are already familiar to you. There are two major pathological processes affecting the vascular tree, atherosclerosis and hypertension. The former, according to the World Health Organization definition, consists of 'focal accumulation of lipids, complex carbohydrates, blood and blood products, fibrous tissue and calcium deposits in the intima associated with medial changes' (*World Health Organization Technical Report*, Series No. 143, 1958). Its main impact is upon large arteries, such as the aorta and its branches, which, in the case of the cerebrovascular tree, are the carotid and vertebrobasilar arteries and the Circle of Willis. It produces its effects by causing stenosis or occlusion of the lumen of the vessel, thereby interfering with blood flow, or by providing a site for the formation of emboli, which may subsequently impact in more distal parts of the cerebrovascular tree.

Hypertension, on the other hand, affects mainly the smaller arteries, producing characteristically thickening of the media, due to hypertrophy of muscle fibres and their replacement by fibrous tissue, and some fibrous thickening of the intima. In addition it gives rise to micro-aneurysms which are found mainly on the striate arteries supplying the basal ganglia and internal capsule and also on the long-penetrating arteries supplying the subcortical white matter. These aneurysms were first described by Charcot and Bouchard (1868), dismissed as artefacts by the majority of pathologists, convincingly demonstrated by Green (1930) and extensively studied by Ross Russell (1963) whose work has been subsequently expanded by Cole and Yates (1967). In brief, it has been shown that the formation of these aneurysms, typically in the sites I have mentioned, is the result of two factors, hypertension and age, and of these two hypertension seems to have the more decisive influence.

These descriptions of the pathological changes produced in atherosclerosis, on the one hand, and hypertension on the other, might lead one to think that, at autopsy at least, if not always during life, subjects who have suffered from these two processes could be clearly separated. This is far from being the case. It is true that there are marked sex and ethnic differences in the prevalence of these two conditions and there is also evidence of a shift in the proportion of vascular disease attributable to them within certain communities (Yates, 1964). Nevertheless, within the individual subject, and even within small groups, it is often difficult to distinguish clearly between the effects of atherosclerosis and the effects of hypertension at autopsy and even more so during life.

The reasons for this are several. Firstly, many subjects suffer from both conditions. Secondly, atherosclerosis may cause hypertension as when it produces a renal artery stenosis, or, according to Dickinson's hypothesis (1965), when it causes hind-brain ischaemia, though the latter I personally doubt. Thirdly, hypertension, though not the primary cause of atheroma, aggravates its formation, producing more plaques in the larger arteries and causing lesions to appear in smaller arteries than are usually affected.

Despite these difficulties I think it is of the utmost importance that we endeavour to separate the effects of these two basically distinct processes, certainly in our thinking, but also at the bedside. Even a cursory reading of the literature shows how often this is not done, the terms cerebrovascular disease, or stroke illness, or, even worse, cerebral insufficiency, being employed as though they referred to a single entity. We cannot hope to make progress unless we distinguish between the various pathological conditions with which we are dealing.

PHYSIOLOGICAL CONCEPTS

Time does not permit me to make more than a very brief mention of the physiological concepts underlying the appraisal and management of cerebrovascular disease, but it is essential to underline one fundamental point, misunderstanding of which has greatly clouded clinical thinking and action. The ultimate factor in maintaining neuronal activity is not blood pressure, but blood flow with the exchange of O_2, CO_2 and metabolites that this latter permits. The relationship between blood pressure and blood flow is given by the simple formula:

$$\text{Cerebral blood flow} = \frac{\text{Blood pressure}}{\text{Cerebrovascular resistance}}$$

From this you can see that a fall in blood pressure would result in a fall in blood flow unless the cerebrovascular resistance also fell proportionately, in which case the blood flow would remain unchanged. This phenomenon of a changing cerebrovascular resistance commensurate with changes in blood pressure so as to keep the blood flow constant is known as auto-regulation. Its existence was hotly denied by many physiologists who failed to demonstrate it experimentally, but the work of Harper (1966) among others has conclusively shown that it does exist. It is a delicate mechanism, readily disturbed, which probably accounts for the difficulty encountered in demonstrating its presence experimentally. The effect of auto-regulation is that mean arterial blood pressure can vary over a range from about 60 to 200 mm Hg. without there being any change in blood flow.

This ability to adjust resistance to maintain flow constant is not the prerogative of healthy arteries. The work of Finnerty, Witken and Fazekas (1954) and Crumpton and Murphy (1952) has shown that the ability is retained despite advancing years and in patients with arteriosclerosis and essential hypertension, though the lower limit of the range of mean arterial blood pressure over which auto-regulation is effective is somewhat higher than in health. It is lost, however, in patients suffering from malignant hypertension, the cerebrovascular resistance remaining high and constant over a wide range of blood pressure.

K

CLINICAL CONSIDERATIONS—'THE HYPERTENSIVE CRISIS'

Fortified by these basic pathological and physiological concepts we can now turn to some clinical applications of this knowledge to problems involving hypertension and cerebrovascular disease. The first of these I want to consider is what may be called the hypertensive crisis. The clinical entity known as hypertensive encephalopathy, which was originally described by Fishberg, in which a sudden rise of blood pressure is accompanied by headaches, vomiting, clouding of consciousness, fits, extensor plantar responses, papilloedema and raised cerebrospinal fluid pressure is well known. The work of Byrom (1954) on the rat has convincingly demonstrated that this is associated with spasm of the cerebral arterioles and the formation of cerebral oedema. Byrom also showed that this may occur on a focal basis and give rise to focal neurological deficits.

The possibility that focal disturbances, giving rise to localized and often transient-clinical deficits, may also occur in man has been insufficiently appreciated. My attention was first called to this by the case which Kendell and I previously reported (1963) of a man of 55 years who, whilst receiving guanethidine for hypertension, began to suffer from episodes of dysarthria, dysphagia and inco-ordination of the limbs, each lasting about 5 minutes. He experienced two or three attacks per day over a period of 10 days. These were thought to be due to drug-induced hypotension, but we were fortunate to be able to record his blood pressure in two attacks, in one of which it rose from 200/115 to 220/150 and in the second to 190/165.

The difficulty in a case of this kind is to be sure that the rise of blood pressure is not the effect of the attack rather than its cause. Since that time I have been fortunate in observing a number of cases in which the blood pressure was being recorded at frequent, regular intervals and was found to rise before the development of the clinical attack. In illustration of this I may quote the case (CV No. 810) of a woman of 64 years who had experienced a number of vertebrobasilar transient ischaemic attacks and who was in hospital being treated with methyl dopa for hypertension. The blood pressure, which was being recorded several times each day, had been maintained at a level of 160/110. Over a period of 48 hours it rose to 220/140, at which stage she developed vertigo, nystagmus and unsteadiness. The blood pressure was reduced to 170-150/110-100 and the symptoms subsided (Marshall 1968). I have also seen it giving rise to transient loss of vision with preservation of the pupillary reflexes similar to the cases described by Jellinek, Painter, Prineas and Ross Russell (1964).

My hypothesis about this type of case is that the arteriolar spasm observed by Byrom in the rat is occurring in a localized area, giving rise to a
focal clinical deficit. You will remember that in my cursory review of the
physiology of cerebral blood flow I mentioned that the one condition in
which the relationship between the cerebrovascular resistance and the
blood pressure is disturbed is malignant hypertension. I think that these
cases are in the early, or pre-malignant, phase of hypertension which is
manifest by focal disturbances of the kind I have described.

It may be asked if some of these incidents may not be due to small
intracerebral haemorrhages in which the effusion of blood is tightly
contained by the cerebral substance. This is certainly a possibility, as we
shall see later, but I find it hard to accept this explanation for repeated
attacks over a short space of time, as for instance occurred in the man I described.

The clinical importance of this condition lies in the fact that many of
these patients are already under treatment for hypertension by hypotensive drugs. The development of a transient, focal, neurological deficit is
often erroneously attributed to hypotension rather than to hypertension
and treatment reduced rather than increased. It is certainly true that focal
neurological deficits can be due to iatrogenic hypotension rather than to
hypertension, but this fact does not justify us making this attribution in the
individual case without firm evidence.

It is often possible to distinguish between these two possibilities by the
circumstances in which the attacks occur. Hypotensive attacks most
commonly occur on change of posture or sudden exertion and are often
accompanied by pre-syncopal symptoms; they are, moreover, frequently
short-lived, being relieved by the patients sitting or lying, as they usually
do when they do not feel well. Hypertensive crises, on the other hand,
frequently develop when the patient is lying down, especially if they are
being treated with drugs which depend for their effect upon the production
of postural hypotension, are slower in evolution and regression, and the
patients commonly show evidence that they are in the pre-malignant
phase of hypertension by the presence of albuminuria and of appropriate
changes in the retinal vessels. If one is fortunate enough to record the
blood pressure during the episode, the diagnosis may be clinched, but it is
surprising how often this observation is neglected even when it is possible
to make it.

Hypertensive crises, even when they produce only focal deficits rather
than diffuse encephalopathy, must be treated with all the vigour that
Pickering, Cranston and Pears (1961) have described for the latter

condition. The need to reduce the blood pressure is urgent and cannot be met by the oral administration of drugs. Intravenous hexamethonium, subcutaneous pentolinium, intramuscular reserpine or some other regime of this kind must be employed until the diastolic blood pressure is at acceptable levels.

MASSIVE CEREBRAL HAEMORRHAGE

Perhaps the most interesting thing to be discussed in relation to massive cerebral haemorrhage is its possible relationship to the microaneurysms I discussed in my brief pathological review. The enormous destruction caused by a massive intracerebral haemorrhage makes it difficult, even with post-mortem injection techniques, radiography and careful serial sectioning of vessels, to determine with certainty the source of a haemorrhage. There is, however, considerable circumstantial evidence to support the view that in many cases it is these microaneurysms. The common dependence of massive cerebral haemorrhage and microanaeurysms upon age and hypertension for their occurrence, the similarity of their distribution in the brain, the occasional case in which a ruptured microaneurysm can be found in the wall of a haemorrhage cavity (Hermann and McGregor, 1940), the more frequent cases in which small haemorrhages are found contained in the perivascular space around an aneurysm, and the ease with which, even under low injection pressures, contrast medium may pass through defects in the wall of the aneurysm into the perivascular space all support this view. This is not to say that every cerebral haemorrhage arises from this source; there are other obvious sources such as berry aneurysms and small arterio-venous malformations. But in the large series examined by Dorothy Russell (1954) over 90 per cent of the cases of cerebral haemorrhage, which were not due to some recognized cause such as a ruptured berry aneurysm, were associated with hypertension. On the basis of the work of Ross Russell and of Cole and Yates it is clear that the brains in these cases would have contained a multiplicity of microaneurysms, hence it is not unreasonable to incriminate them as the source of the haemorrhage.

What can we do in the way of treatment of the massive cerebral haemorrhage? The answer is very little. My unpublished experience of reducing the blood pressure in the acute stage was disappointing, and likewise the experience of McKissock, Richardson and Taylor (1961) of the emergency evacuation of clots showed that it is without value except in the case of haematomata in the posterior fossa (McKissock, Richardson and Walsh, 1960). There are, however, those cases described by Pennybacker (1963) in which an acute onset is followed by an initial stabilization of the clinical condition for about 48 hours after which the patient begins to deteriorate slowly. In some of these cases there are torpedo-shaped, subcortical clots which can be evacuated through a trephine disc with considerable benefit to the patient.

SMALL HAEMORRHAGES FROM MICROANEURYSMS
There is another aspect of the work on microaneurysms which is of considerable interest and that is their possible relationship to transient ischaemic attacks or little strokes. Many hypertensive patients experience focal, neurological deficits such as hemiparesis or dysphasia lasting a few hours or maybe up to a day. Whether one calls these transient ischaemic attacks (TIAs) or little strokes depends upon the time limit one arbitrarily imposes for the duration of the former. Some of these episodes, I have already suggested, are due to focal areas of arteriolar spasm developing in association with a rise in blood pressure. I do not think that this is the cause of them all and I would suggest that some of the remainder may be due to small haemorrhages from microaneurysms, the haemorrhage being contained in the immediate perivascular space.

One is sometimes fortunate enough to obtain convincing clinico-pathological correlation in support of this hypothesis as in the case of a woman (Case No. 43) who developed dysphasia and weakness of the right upper limb without headache, vomiting or loss of consciousness and who recovered completely in 48 hours. One year later she suddenly lost consciousness and died within 24 hours from a massive cerebral haemorrhage in the right hemisphere. There was, however, in the left hemisphere a small old haemorrhage (Prof. W. Blackwood) which seems clearly to have been the cause of her 'little stroke' 12 months previously (Hill, Marshall and Shaw, 1960). If it is true that small leaks from microaneurysms are sometimes the cause of transient ischaemic attacks or little strokes, the clinical implications are obvious. The arbitrary treatment of all cases of TIA with anticoagulant therapy is clearly unwarranted. It is only in the absence of hypertension, and when there is a reasonable probability of the TIAs being due to emboli, that anticoagulants should be used.

The immediate prognosis for the TIAs occurring in the hypertensive subjects is good in that they are likely to recover from the present attack with little or no deficit. This they should be allowed to do, but if, on recovery, the blood pressure remains high—say a diastolic of 110 mm Hg. or above—it should be lowered on a long-term basis.

HYPERTENSION AND CEREBRAL INFARCTION

The relationship between hypertension and cerebral haemorrhage is too well known to require any stress, but the existence of a relationship between hypertension and cerebral infarction comes as a surprise to many. This probably stems from the fact that in the past the term cerebral thrombosis was loosely used in reference to cerebral infarction and thrombosis was erroneously thought to be due to a fall in blood pressure: a double error. In fact, the majority of cases of stroke which are reasonably diagnosed on clinical grounds as being due to cerebral infarction have a blood pressure higher than the average found in subjects of the same age and sex. This was apparent in the series described by Low-Bear and Phear (1961) and in that studied by Prineas and Marshall (1966).

The objection is often raised that the blood pressure at the time the patient has a stroke may not be typical of their usual level; this may certainly be so, but in the series mentioned this was so in only four out of 134 cases and for them the blood pressure recorded when they were convalescent was used in allotting them to their appropriate blood-pressure level. The condition we diagnose clinically as being due to cerebral infarction is, therefore, commonly associated with higher than average blood pressure.

In the study which Prineas and I made of this question we found it rewarding to separate the cases according to the degree to which their blood pressure was elevated. We took the arbitrary level of 110 mm Hg. diastolic as the dividing line and found that the patients separated into two distinct clinical categories. Those patients with a diastolic blood pressure below 110 mm Hg. showed clinical evidence of a large cortical lesion with little recovery, often had a focal disturbance in their EEG, had a high prevalence of bruits in the neck, and at angiography showed a large number of stenotic and occlusive lesions in both the extracranial and intracranial arteries. The patients whose diastolic blood pressure was 110 mm Hg. or above showed clinical evidence of a small, deep-seated lesion which recovered more rapidly, had a normal or diffusely abnormal EEG, a low prevalence of bruits in the neck, and at angiography showed few stenotic or occlusive lesions.

On the basis of this evidence we hypothesized that the former group were infarctions associated predominantly with atherosclerosis causing stenosis or occlusion of large vessels. The function of the small rise of blood pressure in those cases was to aggravate the formation of atherosclerosis rather than to produce the changes of hypertensive vascular disease, namely medial hypertrophy and microaneurysms. The latter group we suggested were due to lesions directly attributable to hypertensive vascular disease.

What is the mechanism of these latter? Some I have already suggested are due to small haemorrhages occurring from the microaneurysms which are present in abundance in patients with hypertension. Others, however, may be associated with thrombosis of the aneurysmal sac which is known to occur, the thrombus spreading to involve the parent vessel and causing an infarct in its territory of supply. Others may be due to occlusion of the penetrating vessel without evidence of aneurysm formation (Miller Fisher, 1965). Whatever the initial cause, the removal of the infarcted material by phagocytes ultimately gives rise to the lacunes which Miller Fisher has studied extensively, both pathologically and with regard to their clinical implications. Some of these lacunes contain haemosiderin, indicating they were the site of a small haemorrhage, but others do not, and the small round cystic cavitation suggests they are the end-result of a small infarct from which the necrotic material has been removed.

It will be remembered that the distribution of the microaneurysms is mainly in the area of the striate and long-penetrating arteries. The clinical picture presented by these cases accords with this localization, their marked tendency to recover with little deficit is compatible with the small size of the lesions and the distribution of the lacunes found at autopsy fits in with the hypothesis that these episodes in patients with considerable hypertension are due to small haemorrhages or infarcts around micro-aneurysms.

The management of these cases suspected of being due to small infarcts around microanaeurysms must be the same as for those in which haemorrhage is thought to be responsible. Our present diagnostic methods are entirely incapable of distinguishing between these two processes, hence it would be folly to think of giving anticoagulants in this situation. The acute episode must be treated by general measures and the blood pressure subsequently lowered on a long-term basis in those cases which require it.

Fortunately the level of the blood pressure and the other features I have described distinguish sharply between these cases and those associated with stenosis or occlusion of large vessels in which endarterectomy or anti-coagulants may have a place. So much is this so that it is usually possible on the basis of the clinical features alone to decide in which cases it will be worth while undertaking angiography.

HYPERTENSION AND TIAS

This may be a convenient point at which to summarize the possible relationships between hypertension and transient disturbances of neurological functions which may be designated as TIAs or little strokes according to their duration. On the evidence available it seems that they may be associated with focal arteriolar spasm, the patients in this situation having very high diastolic pressures of 125 mm Hg. or more which have recently risen, and show evidence of being in the malignant or pre-malignant phase of hypertension. Secondly, some may be associated with small haemorrhages from microaneurysms which are contained by the cerebral substance and do not extend to become massive cerebral haemorrhages. Thirdly, some may be due to thrombosis of these aneurysms and their parent vessel with the resultant formation of a small infarct. Both these latter may ultimately give rise to lacunes, though some of the haemorrhages remain as slits rather than lacunes. The underlying cause in all three processes is hypertension and it is to the control of this that we must look for any advance in therapy.

THE MANAGEMENT OF HYPERTENSION

It is not surprising from that which has been written that it is my contention that the time to treat the cerebrovascular manifestations of hypertension is before they develop. If I am right in thinking that many of the cerebrovascular manifestations of hypertension are associated, in one way or another, with microaneurysms, then there are a number of things we need to know which at present we do not know. For instance, what degree of elevation of the blood pressure must there be to cause aneurysms to develop? How long must the hypertension be present before aneurysms appear? What is the critical age level above which their development is facilitated? When an aneurysm leaks, what determines whether there will be a small, contained haemorrhage or a massive cerebral haemorrhage? The answers to these questions are not beyond our present reach but demand longitudinal studies with close clinico-pathological co-operation on a large scale. Without these answers it will be impossible to ascertain the population at risk and to determine what degree of blood pressure control is necessary to preserve them from that risk.

In the meantime there is considerable evidence concerning the relationship between hypertension, life-expectancy and morbidity in the form of strokes and concerning the effect of lowering the blood pressure upon these. In my original natural history studies I found that among patients who had recovered from a stroke those who had a raised blood pressure had a much poorer life-expectancy than those who did not (Marshall and Shaw, 1959; Marshall and Kaeser, 1961). This observation confirmed the results of other studies (Rankin, 1957; David and Heyman, 1960; Carter, 1963, 1964). The only exception to this has been the study of Merrett and Adams (1966) whose hypertensive patients did not fare worse than the normotensives. The reasons for this different experience are not certain but probably lie firstly in the fact that though their analysis was based on blood pressures recorded in convalescence, they excluded 207 of the 710 patients on the ground that they were not transferred to their care within a month of the onset of their stroke; the effect of an arbitrary exclusion of this kind on a life-table is difficult to predict. Secondly, they took the highest resting pressure that had been recorded; because of diurnal and other variations no figure for blood pressure is completely satisfactory, but among the various possibilities, the highest blood pressure recorded in a series is the least satisfactory because transient rises occur so readily as a result of a variety of stimuli. Thirdly, though an age distribution was not given, the patients were 'elderly'; it is possible the influence of hypertension on mortality is not so great in the older age groups in which other causes of death are more frequently operative.

The effect of lowering the blood pressure of hypertensive patients upon morbidity and mortality has been assessed in a number of studies. The beneficial effect of hypotensive therapy in patients with malignant hypertension (Bjork, Sannerstedt, Angervall and Hood, 1960) and hypertensive heart failure (Smirk, Hamilton, Doyle and McQueen, 1958) has been well documented. The same benefit has been found in patients who showed only raised pressure and Grade 2 retinopathy (Hodge, McQueen and Smirk, 1961). Many of the deaths in untreated hypertensives are due to stroke, the frequency of which is reduced by hypotensive therapy (Leishman, 1963; Hood, Aurell, Falkheden, Olanders and Bjork, 1966; Gifford, 1966).

In patients who had already shown evidence of cerebrovascular disease by having a stroke, I found that long-term hypotensive therapy significantly improved their subsequent life-expectancy (Marshall, 1964). Of course, the therapy must be controlled with care; the diastolic blood pressure must be brought below 110 mm Hg. if the incidence of strokes is to be reduced (Hamilton, Thompson and Wisniewski, 1964), yet it must not be allowed to fall below the level at which auto-regulation will no longer be effective in maintaining adequate flow. Fortunately in the vast majority of patients the margin of safety is considerable and, particularly with modern hypotensive agents which do not have marked postural effects, makes treatment a practical possibility.

Gradually, therefore, the confusion which has surrounded the subject of cerebrovascular disease is giving way to insights as to the pathogenesis of cerebrovascular incidents. These insights are such as to enable us to determine the cause of a stroke in the individual case at the bedside more frequently than in the past. This determination is not as yet so immediately fruitful in terms of therapy as we would like in every case, nevertheless it is an essential step towards rational and effective therapy. I hope this review of the relationship between hypertension and cerebrovascular disease will enable you to approach your cases of cerebrovascular disease with increased interest and confidence.

REFERENCES

BJORK S., SANNERSTEDT R., ANGERVALL G. and HOOD B. (1960) *Acta med. scand.* **166,** 175

BYROM F.B. (1954) *Lancet* **ii,** 201

CARTER A.B. (1963) *Proc. R. Soc. Med.* **56,** 483

CARTER A.B. (1964) In *Cerebral Infarction,* p. 209. Pergamon Press Ltd

CHARCOT J.M. and BOUCHARD C. (1868) *Arch. Physiol. norm. path.* **1,** 110, 645 and 725

COLE F.M. and YATES P.O. (1967) *J. Path. Bact.* **93,** 393

CRUMPTON C.W. and MURPHY Q.R. (1952) *J. clin. Invest.* **31,** 622

DAVID W.J. and HEYMAN A. (1960) *J. chronic Dis.* **11,** 394

DICKINSON C.J. (1965) In *Neurogenic Hypertension,* p. 274. Blackwell Scientific Publications, Oxford

FINNERTY F.A. Jr., WITKIN L. and FAZEKAS J.F. (1954) *J. clin. Invest.* **33,** 1227

FISHER C. MILLER (1965) *Neurology (Minneap.)* **15,** 774

GIFFORD R.W. (1966) In *Cerebral Vascular Diseases. Transactions of Conference held at Princeton, New Jersey, January* 1966, p. 90. Grune & Stratton, New York and London

GREEN F.H.K. (1930) *J. Path. Bact.* **33,** 71

HAMILTON M., THOMPSON E.N. and WISNIEWSKI T.K.M. (1964) *Lancet* **i,** 235

HARPER A.M. (1966) *J. Neurol. Neurosurg. Psychiat.* **29,** 398

HERMAN K. and McGREGOR A.R. (1940) *Brit. med. J.* **1,** 523

HILL A.B., MARSHALL J. and SHAW D.A. (1960) *Q. Jl Med.* **29,** 597

HODGE J.V., McQUEEN E.G. and SMIRK F.H. (1961) *Brit. med. J.* **1,** 1

HOOD B., AURELL M., FALKHEDEN T., OLANDERS S. and BJORK S. (1966) In *Cerebral Vascular Diseases. Transactions of Conference held at Princeton, New Jersey, January* 1966, p. 83. Grune & Stratton, New York and London

JELLINEK E.H., PAINTER M., PRINEAS J. and ROSS RUSSELL R. (1964) *Q. Jl Med.* **33,** 239

KENDELL R.E. and MARSHALL J. (1963) *Brit. med. J.* **2,** 344

LEISHMAN A.W.D. (1963) *Lancet* **i,** 1284

LOW-BEER T. and PHEAR D. (1961) *Lancet* **i,** 1303

McKISSOCK W., RICHARDSON A. and TAYLOR J. (1961) *Lancet* **ii,** 221

McKISSOCK W., RICHARDSON A. and WALSH L. (1960) *Brain* **83,** 1
MARSHALL J. (1964) *Lancet* **i,** 10
MARSHALL J. (1968) *Postgrad. Med. J.* **44,** 42
MARSHALL J. and KAESER A.C. (1961) *Brit. med. J.* **2,** 73
MARSHALL J. and SHAW D.A. (1959) *Brit. med. J.* **1,** 1614
MERRETT J.D. and ADAMS G.F. (1966) *Brit. med. J.* **2,** 802
PENNYBACKER J. (1963) *Proc. R. Soc. Med.* **56,** 487
PICKERING G.W., CRANSTON W.I. and PEARS M.A. (1961) In *The Treatment of Hypertension.* Charles C. Thomas, Springfield, Illinois
PRINEAS J. and MARSHALL J. (1966) *Brit. med. J.* **1,** 14
RANKIN J. (1957) *Scott. med. J.* **2,** 200
ROSS RUSSELL R.W. (1963) *Brain* **86,** 425
RUSSELL D.S. (1954) *Proc. R. Soc. Med.* **47,** 689
SMIRK F.H., HAMILTON M., DOYLE A.E. and McQUEEN E.G. (1958) *Am. J. Cardiol.* **1,** 143
World Health Organization Technical Report (1958), Series No. 143. World Health Organization, Geneva
YATES P.O., (1964) *Lancet* **i,** 65

CHAPTER TEN

THE PATHOGENESIS OF CEREBRAL HAEMORRHAGE AND CEREBRAL INFARCTION

J.R.A.MITCHELL

'Is there any other point to which you would wish to draw my attention?'
'To the curious incident of the dog in the night-time.'
'The dog did nothing in the night-time.'
'That was the curious incident,' remarked Sherlock Holmes.

Ever since John Snow removed the handle of the Broad Street pump, the study of morbidity and mortality statistics has provided us with an epidemiological watch-dog. In respect of coronary heart disease (CHD), the dog has barked loudly and to some purpose and has attracted attention to a number of risk-factors (Table 1). Why, then, has the dog been apparently silent in respect of cerebrovascular disease (CVD) and why has it failed to bring to our notice a list of risk-characteristics for stroke victims? In other ways, the contrast between CHD and CVD has been very striking, for although they both relate to disease of the arteries, they show a very divergent pattern. Thus, within many countries CHD has shown a significant increase with time, while CVD has apparently remained unchanged. Moreover, CHD shows a marked M:F preponderance whereas CVD does not. On an international basis, the two diseases also appear to show dissimilar patterns, in that one can find in Japan a nation with a low CHD and a high CVD mortality rate, and in the United States, a country with an apparently reciprocal pattern.

We should recognize, however, that these patterns have emerged from routine national mortality statistics, and are therefore based on death certificates. It can be argued that documents serve the purpose for which they were designed. A death certificate is a means of reassuring the community that an individual has died of 'natural causes' and it is not necessarily a valid item of scientific epidemiological information. It may reflect, at any given time, the certifying doctor's innate sense of the probability of the various alternative events from which a patient may have died, and his awareness of those death certification labels which are most readily acceptable as being 'natural'. The relevance of this to CVD is

TABLE 1

Risk factors in coronary disease
(based on Oliver, 1966)

Male sex
High blood pressure
Cigarette smoking
Elevated blood lipids
Obesity
Occupation
Impaired glucose tolerance
Genetic factors—blood groups, etc.

TABLE 2

A comparison of mortality rates derived from routine national figures from four countries, with incidence rates for cerebrovascular disease in special studies in these countries (based on Kagan *et al*, 1967)

	Mortality rate (per 1000)	Incidence rate (per 1000)
Japan	2.7	3.6
Norway	0.6	1.3
U.S.A.	0.8	2.2
U.K.	1.1	3.8

shown in Table 2: the death certificate data appear to demonstrate a very wide disparity between Japan and the three other countries. However, when one records the number of new incidents of cerebrovascular disease occurring in groups which were being carefully studied in these countries, the range of values narrows considerably (Kagan, personal communication). In particular, the United Kingdom, which has a mortality rate less than half of that prevailing in Japan, moves up to first place when incidence rates are considered. This suggests that mortality data may not be an accurate indicator of the pattern of CVD in these countries.

In an attempt to measure the inaccuracies inherent in national mortality data, Kagan, Katsuki, Sternby and Vanecek (1967) set out to compare the national figures from Japan, Sweden and Czechoslovakia with figures obtained by special study groups set up in designated areas in these countries. One important preliminary finding was that when they looked at 590 patients in whom fresh cerebral infarction or fresh cerebral haemorrhage had been found at necropsy, they discovered that only half of them had been certified as dying of cerebrovascular accidents. These preliminary results were confirmed by a larger and more detailed series (Table 3),

TABLE 3

A comparison of the clinical diagnosis, as represented by the data on the death certificate, with the necropsy findings (based on Kagan *et al*, 1967)

Accuracy of clinical certification
1478 deaths in hospital

107	where necropsy showed CVD as cause of death 36 of these not clinically diagnosed (20 CH, 4 CI)
44	where necropsy showed CVD as a secondary condition 38 of these not clinically diagnosed
16	where CVD diagnosed clinically, not found at necropsy

which showed that when cerebrovascular disease was thought to be the primary cause of death at necropsy, one-third of the death certificates had not mentioned vascular disease of the nervous system; where fresh haemorrhage or infarction was found at necropsy but was not thought to be the primary cause of death, 86 per cent of the certificates showed no reference to cerebrovascular disease.

Kagan *et al* (1967) went on to show that in those cases where cerebrovascular disease had been specified on the certificate the haemorrhage: infarction ratio was very different from the ratio observed at necropsy. As an amplification of this they compared the results of the three special studies (in Prague, Malmö and Hisayama) with ratios derived from routine national mortality figures for these areas. Table 4 shows their findings and suggests that the national figures grossly overestimate the number of cerebral haemorrhages (CH) and underestimate the number of cerebral infarcts (CI). From the valuable findings of Kagan *et al* (1967) we can hazard a guess at the main sources of error which arise when routine clinical or national data are accepted at face value (Table 5).

From the epidemiological viewpoint, the grouping together of potentially dissimilar diseases effectively obscures the individual behaviour pattern of each of the separate components. If the routine data give grossly

TABLE 4

A comparison of cerebral haemorrhage:cerebral infarction ratios derived from routine national mortality statistics and and from special studies of cerebrovascular disease in the same countries (based on Kagan, 1967)

	CH:CI ratio		
	Routine national 45–54 yrs	Special 50–59 yrs	Routine national 55–64 yrs
Czechoslovakia	6	1.5	5
Japan	12	4.0	7
Sweden	9	2.0	5

TABLE 5

Summary of the main errors which will arise when routine data, based on death certificates, are used in studies on cerebrovascular disease

Main errors
1. Where death is certified without autopsy
2. Where CVD is an associated cause of death

Main results
1. Prevalence figures are erroneous
2. Cerebral haemorrhage:cerebral infarct ratio wrong
3. Total CVD figures conceal differential CH:CI trends

inaccurate figures for total CVD and also falsify the haemorrhage:infarction ratio, can we expect the epidemiological watch-dogs to show us where the risks lie? We must obviously obtain better data, and when acceptable results, based only on autopsy examinations, are studied, quite clear differential behaviour patterns emerge for haemorrhage and infarction.

Yates (1964) showed that in Manchester there had been a decline in autopsy-proven cerebral haemorrhage and a rise in cerebral infarction, giving a dramatic reversal of the ratio of the two conditions (Table 6). This pattern has been confirmed for other countries (World Health Organization, 1965) and suggests that the two conditions have a totally different aetiology and pathogenesis. Although many studies are still in progress in various parts of the world we can summarize the epidemiological situation so far by saying that the studies from Hisayama, Oslo, Framingham and London suggest that cerebral haemorrhage is strongly correlated with raised arterial pressure, and that it is not obviously linked

TABLE 6

Changes in the cerebral infarction: cerebral haemorrhage ratio in autopsies carried out on hospital deaths in Manchester (Yates, 1964)

1877–1900	0.36
1926–37	0.40
1941–46	0.55
1949–50	0.71
1953–54	0.97
1957–58	1.43
1961	1.48

to cardiac infarction, limb gangrene or other manifestations of large artery obstruction; cerebral infarction, on the other hand, is less strongly correlated with hypertension but is much more closely linked with cardiac infarction and with limb artery disease (Kagan, personal communication). Does the pathological evidence provide a supporting structure for this epidemiological data and an explanation for the differential behaviour of cerebral haemorrhage and cerebral infarction?

Cerebral haemorrhage can be dramatic and has always had an aura of mystery and irrevocability; the drama and the mystery were expressed very clearly in Wepfer's *Historiae Apoplecticorum* (1658). In describing the history of a 45-year-old man he noted '. . . In the year 1655, the seventh day of November, the fifth day of the full moon, in the morning sane and sound, he did much of everything; he assisted the Most Reverend

Lord Abbot in the carrying out of the sacraments, accomplishing which, he gathered up things in the dining room, according to his custom; the Abbot by chance wished to decide with him the fate of the servants, found him prostrate upon the ground, insensible to shouts, to shaking and pinching of the body, the same in the trunk senseless; there was hurry, antiapoplectic water given and other things which were at hand were tried, but all in vain and I was summoned. I arrived in about half an hour, I saw him livid from pallor, deprived of all sensation and animate motion, with his nostrils cold to the touch. His pulse at first strong, full, quick, soon afterwards weaker, smaller and more frequent, his breathing also more laborious, soon it became irregular, and many times it appeared about to cease from within.' Some hours later the patient died and Wepfer described the necropsy findings: '. . . I opened the head; the skull removed and the dura mater being cut into pieces much blood flowed from the space, which is very roomy between this and the thin meninges, copiously that is, from all sides and everywhere it poured forth; nor truly had the blood collected solely about the base of the brain, but covered it all over to the top anteriorly and posteriorly, indeed it had forced itself into nearly all the windings of the brain, as many as there are. The ventricles laid open I found them all filled up with blood; but a portion of them excepted, the lateral ones, for the floor was badly torn as it were, as if the fissures were stretched apart by too great a quantity of blood. I was able without difficulty to estimate the weight of the quantity of extravasated blood to have totalled two pounds. The whole brain ventricles and surface were contaminated by blood in large amounts and crumbly; there was nothing further to observe, I was able to find no ruptured vein or artery.' The source of the bleeding in primary intracerebral haemorrhage continued to puzzle pathologists for three centuries. In 1868 Charcot and Bouchard allowed the brain to digest itself away from the blood vessels in cases of cerebral haemorrhage and found some aneurysmal swellings on the intra-cerebral arteries. Subsequent workers, however, showed that the Charcot-Bouchard lesions were not aneurysms but were perivascular haematomas and by one of the swings of opinion which bedevil scientific work, it was concluded that once these lesions had been shown to be 'false aneurysms', no other abnormalities need be sought. By laborious dissection, Green (1930) found several true intracerebral aneurysms and in 1963 Ross Russell adapted to the cerebral vasculature the injection-micro-radiography techniques used in coronary artery studies by Schwartz and Mitchell (1962a) and found large numbers of aneurysms (Figs. 1 and 2). They were especially common in the basal nuclei, and bore an unequivocal relationship to elevated arterial pressure (Table 7). Cole and Yates (1967) have recently applied a similar technique to 100 patients with in-life diastolic blood-pressure levels of over 110 mm Hg and with large hearts at necropsy. They also studied 100 age- and sex-matched patients who did not

TABLE 7

Prevalence of intracerebral aneurysms in patients with raised blood pressure
(Ross-Russell, 1963)

	Prevalence of aneurysms, expressed as numbers of patients with stated number				
	None	*1 to 5*	*6 to 10*	*More than 10*	*Unsatisfactory injection*
Control group	25	7	2	1	3
Hypertensive group	1	5	5	4	1

TABLE 8

The prevalence and main characteristics of intracerebral aneurysms (Cole and Yates, 1967)

	100 patients (DBP over 110. Increased heart weight)	*100 age- and sex- matched controls*
Cerebral haemorrhage	20	1
Aneurysms	46% (86% in those with haemorrhage) (35% in those without)	7%
Youngest aneurysm subject	44 Only 2 under 50	66
Sex ratio in aneurysm subjects	M = F	

fulfil these critera (for brevity I shall refer to these two groups as 'hypertensives' and 'controls'). Table 8 summarizes some of the findings: as one might expect, the majority of cerebral haemorrhages occurred in the hypertensive group, and the prevalence of aneurysms was clearly correlated with both the height of the blood pressure and with the presence of haemorrhage. A most important point was that the aneurysms were age-related, and occurred at a younger age in the 'hypertensive' group. In

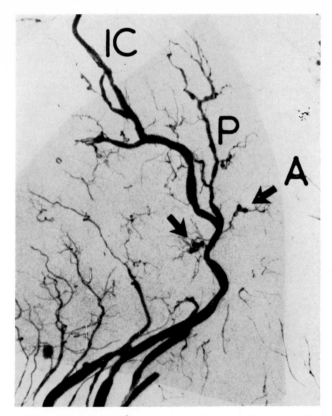

FIG. 1. Micro-radiograph of arteries in coronal section of basal ganglia from patient with raised arterial pressure. Note the small aneurysms. From Ross-Russell, 1963 (by kind permission).

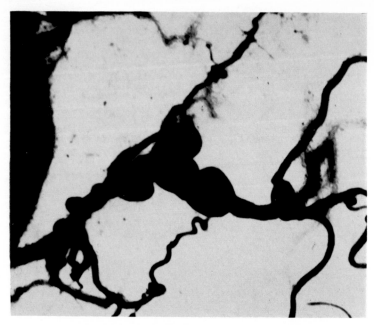

FIG. 2. Enlarged view of area marked A on Fig. 1. From Ross-Russell, 1963 (by kind permission).

contrast to obstructive arterial disease, which I shall discuss later, there was no sex difference in aneurysm prevalence. Table 9 shows the distribution of the lesions, which parallels closely the frequency with which intracerebral haemorrhage occurs in the various sites. Cole and Yates (1967) showed that medial thinning and loss of elastica were the characteristic findings at the point of origin of the aneurysms and although they stress that the relationship between the lesions and cerebral haemorrhage is not yet known, they conclude that aneurysms develop with advancing age in a proportion of the population, and that they do so at an earlier age and with a greater frequency in patients with raised arterial pressure.

Two important questions then arise. First, why has the prevalence of cerebral haemorrhage been declining over the years (Yates, 1964) if it is

TABLE 9

The distribution, number and size of intracerebral aneurysms (Cole and Yates, 1967)

Site	Cerebral hemispheres (most often in basal ganglia and subcortical white matter)	53
	Pons	15
	Cerebellum	4
Number	9–80 per affected brain (15–25 usually)	
Size	0.05–2.0 mm	

casually linked to hypertension which, as far as we know, has not become less common? Perhaps the age-factor, stressed by Cole and Yates (1967), provides a clue, for as Mainland (1953) showed, fatality-rates are competitive (Mitchell and Schwartz, 1965). If a condition which is very closely linked to high blood pressure can only express itself after a considerable number of years, can it be pre-empted by a condition such as cardiac infarction which is less closely linked to high blood pressure but which can occur at an earlier age and which is becoming more common? The second question is a therapeutic one. Will reduction of elevated arterial pressure levels prevent the initial aneurysm formation and will it prevent aneurysms which have already formed from giving rise to intracerebral bleeding? There is already evidence from long-term clinical studies (Leishman, 1959, 1963) that this may be so, and these data provide another good reason for not accepting at face value that 'benign' hypertension is aptly named.

The other major form of cerebrovascular disease is cerebral infarction. Although Morgagni (1761) drew a clear distinction between 'sanguinous' and 'non-sanguinous' strokes the supposed cause of cerebral softening has shown a characteristic cyclical pattern of discovery, oblivion and rediscovery. Abercrombie (1828) was in no doubt that arterial obstruction was an important factor and likened cerebral softening to limb gangrene. Although Todd (1844) stressed the importance of neck artery occlusion in cerebral infarction, during the remainder of that century attention was increasingly focused on the intracranial arteries. An infinite variety of neurological patterns was defined and each pattern was supposed to relate to obstruction of a minute and often eponymous artery. Although the clinical syndromes (i.e. areas of infarction) were accurately demarcated, we now must accept that many of the arterial obstructions exist in name only. Thus the syndrome of 'posterior inferior cerebellar artery thrombosis' described by Taylor (1871) and Wallenberg (1895) is more often due to vertebro-basilar occlusion than to a recognizable thrombus in the artery from which it takes its name.

Early this century there was a reawakening of interest in the neck arteries (Ramsay Hunt, 1914). This was short lived, but the pioneer pathological studies of Hutchinson and Yates (1957) and the parallel developments in angiography and neuro-surgery, produced a lasting effect. There is little doubt that Hutchinson and Yates's early studies (1957) failed to take account of the prevalence of neck artery disease in the general population (Schwartz and Mitchell, 1961) and of the minimal effect which anatomically severe neck artery stenosis may have on cerebral blood flow (Brice, Dowsett and Lowe, 1964). There was therefore an undue emphasis given to neck artery narrowing by clinical investigators and a corresponding neglect of the equally important role of neck artery disease as a source of emboli and as a basis for subsequent complete thrombotic occlusion.

A recent study by Battacharji, Hutchinson and McCall (1967) provides important evidence on the role of these latter processes. They studied fifty-seven patients with cerebral infarction and eighty-eight patients without infarction. There were 130 infarcts in the former group (21 per cent cortical, 51 per cent subcortical and 28 per cent involving both zones). One hundred and seven of them were in the carotid territory and twenty-three in the basilar area. There was a somewhat higher prevalence of neck and intracranial artery stenosis in the infarct group, but the stenosis alone cannot have been the critical factor for although the bulk of the infarcts were in the middle cerebral territory, there was little to choose between the anterior, middle and posterior cerebral arteries in severity of stenosis.

In the coronary tree, Mitchell and Schwartz (1963) showed that the vital factor which differentiated patients with and without infarction was not just the severity of stenosis but whether an occluding thrombus was present. Battacharji, Hutchinson and McCall (1967) found that a similar relationship applied in cerebral infarction; while there were 64 carotid occlusions (42 R and 22 L), 30 vertebro-basilar occlusions (16 R, 14 L) and 28 intracranial occlusions (middle cerebral 11, posterior cerebral 8) in their infarct group, they found no occlusions in their control group.

The problems of cerebral infarction are therefore similar to those of artery wall disease and luminal occlusion in other territories. Some pathologists have suggested that each territory is a law unto itself and that disease

TABLE 10

Amount of aortic disease in age-matched groups of men with and without carotid and vertebral artery stenosis

Group	Percentage of aortic surface affected		
	Fatty streaks	*All raised lesions*	*Complicated lesions*
No carotid stenosis	11.7	24.6	7.8
Carotid stenosis	9.9	36.3	16.4
No vertebral stenosis	13.0	24.1	7.9
Vertebral stenosis	9.1	34.1	14.8

severity in one field does not correlate with that in other areas. In part this has arisen because all forms of plaques have been considered together under some generic heading such as 'atherosclerosis'. As fatty streaks bear a doubtful evolutionary relationship to other types of lesion, and are of no immediate clinical significance, their inclusion in inter-territorial comparisons serves to blur the picture. When one compares like with like, there is a significant correlation between stenosing plaques in the coronary arteries, neck arteries, leg arteries and aorta (Mitchell and Schwartz, 1962) and Table 10 shows the relationship between aortic plaques and neck artery stenosis.

Although arterial plaque formation tends to progress in a parallel fashion in the various areas, we must also recognize that it is a localized and not a generalized disease, for there are sites of preferential attack and sites of relative freedom from disease. For example, Schwartz and Mitchell

(1962b) drew attention to the gross involvement of the internal iliac arteries and to the sparing of the external iliac arteries. Any purely systemic theory of causation (whether implicating dietary fat, smoking, sucrose consumption, hypertension, or stress) cannot account for the localization of plaques, which is also seen to advantage in the neck arteries (Schwartz and Mitchell, 1961) where the carotid sinus is a 'bad' segment, whereas the cervical part of the artery, a few centimetres distal to it, is a 'good' segment.

We have already seen that in the causation of cerebral infarction, the thrombotic process is more critical than the wall disease. In recent years, considerable progress has been made in the slow struggle to remind contemporary workers that the nineteenth-century pathologists were accurate observers, who had shown that an arterial thrombus is totally different from a blood clot (Poole and French, 1916). In view of this structural difference there was no *a priori* reason to suppose that the study of blood coagulation would clarify the causation of thrombosis or that the giving of anticoagulant drugs would prevent it. Half a century of abortive study has indeed served to confirm these points.

Once the key role of platelets in thrombus formation was recognized increasing attention has been given to the behaviour of these cells, in an attempt to identify a characteristic and abnormal pattern in patients with thrombotic disease. Enhanced platelet adhesiveness to glass has been shown in patients with cerebral infarction (Millar and Dalby, 1965). Such enhancement is, however, entirely non-specific (Mitchell, 1967), being found in other neurological diseases such as disseminated sclerosis (Caspary, Prineas, Miller and Field, 1965; Wright, Thompson and Zilkha, 1965) and in patients with cerebral tumours (Millac, 1967). It may therefore be a secondary effect produced by destruction of nervous tissue. A more specific characteristic has recently been outlined by Hampton and Mitchell (1966). This pattern (a dissociated sensitivity to adenosine diphosphate and to noradrenaline in electrophoretic mobility studies) occurs in patients with ischaemic heart disease, hypercholesterolaemia and peripheral vascular disease. It is found in patients with cerebral infarction but not in patients with cerebral haemorrhage, and it appears to be due to the presence in the blood of a particular phospholipid (a low-density lipoprotein lecithin) from which lysolecithin can be released by a labile enzyme (Bolton, Hampton and Mitchell, 1967). The precise significance of this has yet to be established but the recognition of therapeutic manœuvres which will change this pattern might well be a very significant advance, and might be of long-term, prophylactic significance. In the short term, however, we urgently need an effective anti-thrombotic agent. Conventional anticoagulants do not affect platelet behaviour, and do not modify the natural history of arterial thrombosis. We do not yet know whether some of the agents which affect platelet response *in vitro* and *in vivo* will prove to have anti-thrombotic activity (Hampton, Harrison, Honour and Mitchell, 1967). Because of its natural origin and its phenomenal potency, the agent we are currently most interested in is the prostaglandin PGE1 (Emmons, Hampton, Harrison, Honour and Mitchell, 1967) and we have now collected evidence that when it is given intravenously to man, platelet behaviour is markedly modified. Further studies are in progress to assess whether it can be used in the treatment of vascular occlusion.

In summary, it has been my contention that the epidemiological watchdogs have been barking very loudly and clearly in CVD, but that there has been so much extraneous noise that their message has been unheeded. The sources of this noise, which we must try to silence, have been described. The actions required to deal with the real message are simple and are based on a clearer understanding of the different pathogenic processes in cerebral haemorrhage and infarction. First we must determine how raised arterial pressure produces the intracerebral micro-aneurysms, what factors govern their rupture and whether controlling the hypertension will prevent their formation and disruption. This is a problem unique to the central nervous system. Second, we must determine how and why thrombi form and we must develop agents which will modify this process. This is a problem which is common to other critical territories such as the heart and limbs and is part of the general problem of obliterative arterial disease.

REFERENCES

ABERCROMBIE J. (1828) In *Pathological and Practical Researches on Disease of the Brain and the Spinal Cord.* Waugh & Innes, Edinburgh
BATTACHARJI S.K., HUTCHINSON E.C. and McCALL A.J. (1967) *Brit. med. J.* **3,** 270
BOLTON C.H., HAMPTON J.R. and MITCHELL J.R.A. (1967) *Lancet* **ii,** 1101
BRICE J.G., DOWSETT D.J. and LOWE R.D. (1964) *Lancet* **i,** 84
CASPARY E.A., PRINEAS J., MILLER H. and FIELD E.J. (1965) *Lancet* **ii,** 1108
CHARCOT J.M. and BOUCHARD C. (1868) *Arch. physiol. norm. et path.* **1,** 643 and 725
COLE F.M. and YATES P.O. (1967) *J. Path. Bact.* **93,** 393
EMMONS P.R., HAMPTON J.R., HARRISON M.J.G., HONOUR A.J. and MITCHELL J.R.A. (1967) *Brit. med. J.* **2,** 468
GREEN F.H.K. (1930) *J. Path. Bact.* **33,** 71
HAMPTON J.R., HARRISON M.J.G., HONOUR A.J. and MITCHELL J.R.A. (1967) *Cardiovasc. Res.* **1,** 101

HAMPTON J.R. and MITCHELL J.R.A. (1966) *Lancet* **ii,** 764

HUNT J.R. (1914) *Am. J. med. Sci.* **147,** 704

HUTCHINSON E.C. and YATES P.O. (1957) *Lancet* **i,** 2

KAGAN A., KATSUKI S., STERNBY N. and VANECEK R. (1967) *Bull. Wld Hlth Org.* Vol. 36

LEISHMAN A.W.D. (1959) *Brit. med. J.* **1,** 1361

LEISHMAN A.W.D. (1963) *Lancet* **i,** 1284

MAINLAND D. (1953) *Am. Heart J.* **45,** 644

MITCHELL J.R.A. (1967) Platelets and thrombosis. In *Annual Rev. Scientific Basis of Medicine.* The Athlone Press, London

MITCHELL J.R.A. and SCHWARTZ C.J. (1962) *Brit. med. J.* **1,** 1293

MITCHELL J.R.A. and SCHWARTZ C.J. (1963) *Brit. Heart J.* **25,** 1

MITCHELL J.R.A. and SCHWARTZ C.J. (1965) In *Arterial Disease.* Blackwell Scientific Publications, Oxford

MILLAC P. (1967) *Brit. med. J.* **4,** 25

MILLAR J.H.D. and DALBY A.M. (1965) *Proceedings of the 8th International Congress of Neurology, Vienna*

MORGAGNI J.B. (1761) *De Sedibus et causis morborum.* Trans. ALEXANDER B., 1769. Millar & Cadell, London

OLIVER M.F. (1966) *Proc. R. Soc. Med.* **59,** 1180

POOLE J.C.F. and FRENCH J.E. (1961) *J. Atheroscler. Res.* **1,** 251

ROSS RUSSELL R.W. (1963) *Brain* **86,** 425

SCHWARTZ C.J. and MITCHELL J.R.A. (1961) *Brit. med. J.* **2,** 1057

SCHWARTZ C.J. and MITCHELL J.R.A. (1962a) *Brit. Heart J.* **24,** 761

SCHWARTZ C.J. and MITCHELL J.R.A. (1962b) *Circ. Res.* **11,** 63

TAYLOR H. (1871) *Brit. med. J.* **2,** 527

TODD R.B. (1844) *Med. chir. Trans.* **27,** 301

WALLENBERG A. (1895) *Arch. psychiat. nervKrankh.* **27,** 504

WEPFER J.J. (1658) *Historiae Apoplecticorum.* Jassonio-Waesbergios, Amsterdam, 1724

WORLD HEALTH ORGANIZATION (1965) *Epidemiological Vital Statistics Reports* **18,** 253

WRIGHT H.P., THOMPSON R.H.S. and ZILKHA K.J. (1965) *Lancet* **ii,** 1109

YATES P.O. (1964) *Lancet* **i,** 65

CHAPTER ELEVEN

NUCLEIC ACID METABOLISM IN NEUROLOGY

W. RITCHIE RUSSELL

The development during the past two decades of what is generally referred to as molecular biology has influenced to a startling extent a great many biological sciences. The main concern of these studies has been with the nucleic acids, and virologists have led the way in many aspects of this work.

As Schmitt (1966) recently explained, these investigations have been chiefly concerned with a study of cellular and subcellular interaction by which genetic or immunological information can be stored, transferred and read out or retrieved in macromolecular DNA, RNA and protein polymers.

The importance of these advances to the neurological and psychological sciences depends on the widely held view that these new genetic and immunological discoveries may provide important clues with regard to the processes involved in memory in the neurological sense.

Every cell contains and maintains a genetic code for the whole body, part of which under certain circumstances transmits the structure and functions of the cells in each organ or part of an organ. These are intricate and precise methods of transferring information and therefore in one sense constitute a type of biological memory.

Professor Campbell will explain this in Chapter Twelve and my task is to try and indicate the importance of this subject for some of the neurological sciences.

From the point of view of neurology there are two special aspects of this knowledge of genetic coding. The first is concerned with the elucidation of disorders of nerve cell metabolism whether inherited or acquired. These faults in cell metabolism may result respectively in a failure of normal brain activity to develop after birth, or in later life the brain mechanisms may degenerate and thus lead to a progressive dementia.

Conditions such as phenylketonuria, amaurotic family idiocy, mongolism and Huntington's chorea are relevant to these problems. The second aspect of this new knowledge is concerned intimately with the normal physiology of nerve cells and requires attention in detail, for this is our chief concern here.

As Melton (1963) has emphasized, the problems of memory have in the past lacked an interdisciplinary research interest, but things have changed greatly and the students of molecular biology, biochemistry, neurophysiology and psychology are busy trying to understand each other's language.

The normal physiology of nerve cells is concerned with many activities. One of these is the transmission of 'all or nothing' impulses conveying information over a distance. Another type of activity occurs at either end of the neurone where, on the one hand, impulses arise in the soma-dendrite-initial segment and, on the other, impulses are transformed by synaptic transmission into post-synaptic potentials. These post-synaptic potentials, either excitatory or inhibitory, differ in most of their characteristics from the all or nothing impulse which serves as their intermediary. For instance, they are graded with variable amplitudes and signs so that, in some cases at least, whether or not an impulse fires off down the nerve depends on the spatial and temporal integration or algebraic summation of these slower potentials. These areas at either end of the neurone are, most probably, important sites of plasticity and learning in the nervous system. Disuse, for example, has been shown to decrease the ease of transmission through the affected synapses and tetanization to facilitate transmission through disused and, to a lesser extent, normal synapses. These changes persist for seconds, minutes and perhaps hours. Such a time course, though extremely long from the physiologist's viewpoint, is much briefer than neurologists and psychologists usually deal with. Another example of plasticity is illustrated by experimental chromatolysis of anterior horn cells which results, not only in the classical 'structural' changes in RNA, but also in striking functional changes in thresholds of dendrite and initial segment membranes. Thereupon a previously simple monosynaptic reflex assumes a complex polysynaptic appearance. Theoretically at least, less dramatic changes in RNA metabolism effected by other means may be associated with equally significant functional changes. The changes in firing of a single neurone are but part of changes in patterns of firing of populations of neurones, but physiology has not yet been as successful in studying populations and natural patterns as it has in the study of single neurones and their responses to artificially produced highly synchronous patterns of afferent bombardment. It seems fair, therefore, to say that the physiology of highly complex and patterned activity as well as that of long-term CNS changes, both crucially important in any discussion of memory, have yet to be elucidated.

The problems regarding the physiology of memory and learning have recently become a centre of interest following suggestions that the coding capacity of RNA may play an important part in the story.

Thus Hydén (1966), who has led this field of thought since 1959, proposes a memory hypothesis based on the capacity for nerve cells to synthesise RNA and protein. A new form of RNA with a base sequence would be coded to correspond to the pattern of electrical activity reaching the cell.

It has also been argued that the patterns of behaviour transmitted by the chromosomes might be similar to the memory evident after learning and experience. Fairly complex behavioural patterns can certainly be inherited and therefore coded in the DNA, but the genetic mechanisms seem to operate more by influencing the pattern and growth of neural mechanisms than by anything comparable to the effects of learning and experience. Further, as Richter (1966) has emphasized, such neuronal differentiation takes place relatively slowly and thus offers no parallel to the almost instant capacity to recall a visual image or a motor skill.

It is reasonable to consider that the special metabolism of the nerve cell which has been demonstrated might be entirely required for, and concerned with, the unique capacity of the nerve cell to maintain an irritable membrane which can be depolarized at a great variety of speeds up to hundreds of times a second. The transfer of normal or abnormal characteristics through the genetic code may certainly in itself be looked on as a type of memory, but one which is very different from the plastic changes which must take place in brain cells in relation to the development of patterns of behaviour and of learning.

It seems that neuronal activity is associated with a rapidly changing production of proteins with the RNA as an activator, so that activity of any one part of the brain leads to a demonstrable increase in RNA activity.

When first made aware of this field of research and the suggestions regarding memory and learning, the general theme seemed to be unattractive, for the early ideas suggested complex information being stored in single cells—a conception which is very foreign to current views which insist that even the simplest aspect of brain activity involves a vast number of interconnecting units. However, if through the complex capacities of RNA coding a pattern of afferents arriving at the cell could be selected to determine the neurone's reaction as regards speed of discharge and timing of response, then this would be most exciting and we might really begin to understand how the CNS works.

Here it may be helpful to refer to some of what is known about memory mechanisms in relation to the nervous system. It is perhaps necessary to reiterate that the activity of a nerve cell leads to repetitive discharges through its axon which generally pass to activate one or more other nerve cells. The strength of neuronal activity is based largely on the frequency of discharges via its axon. Even when apparently at rest the nerve cell may discharge spontaneously from time to time. This is a point of great theoretical interest, for the spontaneous unconscious discharge of neurones throughout life may play a part in maintaining patterns of communication between cells and may thus help to maintain memories.

If we consider, for example, the effects of a simple visual experience, this seems to activate a vast number of neurones in the brain at a very high repetitive rate of discharge over a considerable period; and this process in the trained brain involves complex grades of recognition and comparisons with previous visual experiences. There is also an emotional association of, say, pleasure or fear according also to previous experience: high arousal may facilitate learning, as measured by delay recall experiments (Klein, Smith and Kaplan, 1963). Even if the new experience is of no interest, it is generally held and can therefore be recalled for a few seconds, but then, in the absence of any special alerting activity, it begins to fade quickly. However, if the visual experience caused sufficient alerting to encourage remembering for, say, a few hours, then it seems that a repetitive process is set in operation which with the passage of time establishes a memory trace which can often be recalled perhaps a week or more later. There is now abundant evidence that there must be an unconscious ongoing process of establishing a memory which is a very vital aspect of the process. This is put out of action very promptly in its earliest stages if the brain is inactivated by cerebral concussion or by electric convulsion occurring within a few seconds of the visual experience. As the minutes pass, the memory becomes more firmly established but even after some hours the recently established memory is relatively vulnerable to cerebral insult. The entirely unconscious progressive establishment of a memory is sometimes thought of as depending on a reverberating circuit of communication between millions of neurones repeating the pattern of activity stimulated by the visual experience, and maintained on the basis of a short-term neural change (Milner, 1957).

The clinical study of retrograde amnesia (RA) in relation to cerebral concussion provides adequate proof of the importance of the time factor in relation to the development of a memory trace (Russell and Newcombe, 1966), and these observations are fully supported by observing the occurrence of similar periods of RA in relation to electroshock therapy (Williams, 1950) and also in relation to the development of Korsakoff's syndrome in which a long period of RA is a constant finding.

The time factor concerned with establishing a memory trace has been studied in animals by administering electroshock at various intervals after training runs. Several studies have demonstrated interference with learning on recall, if shocks are given within 10 seconds of the event to be recalled (Chorover and Schiller, 1965; Gerard, 1961). It seems likely, therefore, that the process of consolidation is impaired by the ECT *per se*, and conversely that the neural changes necessary for learning may require to develop over a period of time.

There are also many observations regarding the relative vulnerability of recent memories and the great strength of some remote memories. These observations have been made during and after recovery from severe head injuries (Russell, 1959), and also in relation to the deteriorations of old age (Welford, 1958).

This capacity to add to the memory store only operates, however, if primitive structures in the temporal lobes of the brain and diencephalon (in the limbic system) are in a healthy state. When these are diseased, as in Korsakoff's syndrome (v.i.), new memories cannot be formed. Recent knowledge regarding the activating effects of the reticular formation of the brain stem have taught us to think of one part of the brain influencing the behaviour of another, and it looks as though the limbic system plays an essential role in enabling other parts of the brain to store new information.

The reason for repeating to you these well-established features of the simplest type of remembering is to emphasize again the vast system of intercellular communication which is involved in a very simple item of CNS activity. It seems possible that this system repeats over and over again until some change takes place which establishes the original visual pattern so strongly that it becomes available for immediate recall and for innumerable correlations with future visual information arising for analysis.

In general, the capacity to develop this simple memory is thought to depend on repetitive activity, which is probably maintained by minor physico-chemical changes at synapses which depend on activity resulting in a lower threshold. These changes strengthen as the minutes and hours pass and on this basis it is easy to understand the relative vulnerability of recent memory after concussion or convulsive therapy. However, this process works better if there are no distractions after the material to be remembered. Many of you will know from experience that reading before sleep is an aid to learning and there are many psychological studies concerned with the interference with memory by brain activity following the episode to be retained. It has also been reported that sensory deprivation may improve the retention and recall of learned material (Grissom, Snedfeld and Vernon, 1962).

Further, an event that is unique in one's experience is relatively easy to remember, but if the unique event subsequently becomes commonplace, then the original occurrence (once unique) may become lost to recall.

In order to study the physiology of memory and learning it is probably best to confine one's attention to some simple facet of this fundamental capacity, and therefore it is with hesitation that I continue to refer to some other aspects of memory lest these may confuse the issue by their irrelevance. However, it is wise to have some idea of the complexities we have to deal with.

The first of these is concerned with the way in which a trained brain can suddenly be alerted by a highly complex combination of circumstances. The exhausted mother who is wakened only by the cry of her child is a remarkable example, as is the reaction of the watch-dog to an unfamiliar footstep; or the experimental subject, in selective attention experiments, who normally can ignore the second channel is nevertheless able to perceive certain signals in it, for example his own name, with lowered thresholds (Triesman, 1964). As individuals we are all to a variable degree liable to be suddenly roused to anger or apprehension by a highly complex series of occurrences. Nowadays we may say this is due to the alerting effects of the reticular formation in the brain stem, but the point is that the circumstances of alerting are so complex in relation to previous experience and learning that they must require the complexities of the cerebral hemispheres for their recognition. Thus, there must be a mechanism highly facilitated whereby complex associations in the CNS can jump quickly to alert the whole mechanism—a hot line to the reticular formation!

The next point which requires discussion is concerned with Korsakoff's syndrome, for this remarkable condition illustrates some important characteristics of remote memory. In this syndrome all capacity to add new information to the memory store seems to be lost. Immediate repetition of a series of digits is possible, but within 10 minutes there is no recollection of the digits, of the test being done, or of the examiner even having been seen before. This syndrome develops in relation to some very precise anatomical defects developing in the brain, all of which are concerned with bilateral involvement of the limbic system. Thus, disease or excision of the hippocampus in both temporal lobes (Milner and Penfield, 1956), of the corporal mamillaria, of the medical dorsal nucleus of the thalamus (Adams, Collins and Victor, 1961), or of structures near the third ventricle (Williams and Pennybacker, 1954; Victor and Yakolef, 1955) all produce Korsakoff's syndrome. This condition can only be studied properly in the adult (trained) brain and in such cases there is often a remarkable preservation of remote memories, skills and intellectual abilities which were developed before the illness or injury. As Talland (1965) pointed out, *time sense* in these patients has lost its cues; initiative is lacking, and there is a failure to activate and sustain a search for the retrieval of memories. As the last memories to be recalled are of events which occurred some months before the illness or injury (owing to the retrograde amnesia), it is inevitable that attempts to converse about current affairs are confused, repetitive and involved in confabulation. The most startling variety of this syndrome is that which occurs in tuberculous meningitis, for here a Korsakoff state may continue for months or even years (Williams and Smith, 1954), and yet after cure of the meningitis a good recovery of the capacity to learn may occur although of course there remains amnesia for the whole illness.

It is, of course, highly significant that during this amnesic state the remote memories which can be readily recalled were established with the aid of the limbic system which is now out of action or destroyed—hence the conception of the limbic system as driving the rest of the CNS to hold information, and also the view that old memories are maintained by well-developed structural changes in many parts of the cerebrum remote from the limbic system.

There is a complexity here, for the physiology of recalling is highly obscure and yet this is also concerned somehow or other with temporal lobe function. Recalling not only involves remembering a previous event but also a strong sensation of familiarity which enables the individual to recognize the right answer. The *dejà vûe* sensation of the temporal lobe fit caricatures this function while the artificially induced hallucination on temporal lobe stimulation illustrates a distorted recalling phenomenon. However, in Korsakoff's syndrome recalling of remote events may be remarkably accurate and it is not at all clear how this can happen after temporal lobe excision. However, the patient lacks the normal sense of correctness or familiarity in relation to what he remembers, for he may insist very vehemently that his confabulations are as true as his true memories.

Bickford and his associates (1958) have reported that stimulation of the temporal lobes (as a prelude to surgical excision), provoked a reversible retrograde amnesia. The stimulation thus appeared to affect the retrieval, rather than the consolidation process; and the retrograde amnesia was shown to be a function of the length of stimulation, covering a wide time scale, from 1 second to several weeks. These observations are comparable to the shrinkage of RA during recovery from cerebral concussion, for much of the RA first observed recovers, leaving a short period for which the RA is permanent. There is therefore a storing and also recalling system which are closely linked and yet are also in some respects independent: this is a very mysterious aspect of our subject.

These, then, are a very few of the observations being made on the various factors which influence memory and learning, and there is little doubt that great advances in the understanding of these problems will follow before long.

In conclusion I should like to emphasize some aspects of the problem.

1   From the point of view of neurophysiology an experience which continues for a second or two allows time for many hundreds of repetitive neuronal discharges.

2   These start a process which in the course of some minutes may establish a firm memory trace which holds some of the original experience for future recall.

3   Shortly after this process starts, the receiving areas used are 'cleared' for the reception of new information, so the consolidating process is not maintained by the receiving cells.

4   The so-called on-going process of consolidation occurs unconsciously and if allowed to continue undisturbed in a healthy brain will within a few minutes establish a memory trace which is resistant to electroshock, cerebral concussion or hypothermia.

5   During this period of consolidation, some physical changes must operate in relation presumably to the neuronal networks involved in relation to the original experience.

6   This pattern could presumably be maintained by repetitions from spontaneous neuronal activity, and this might cause physico-chemical changes at synapses which would allow a pattern to be resumed after interruption by, say, hypothermia.

7   It is tempting to consider the earlier processes of establishing an engram to be physical, and that at some stage of the process the molecular biology of the nerve cell bodies should take control, but there is a grading in the vulnerability of memories which seems to stretch out (back) indefinitely through the years so that we must, I think, look for one mechanism for all stages of consolidation, or if there are two mechanisms, they must be mutually interdependent.

8   If an original experience involves a vast number of neurones, it seems probable that the recall of that experience will activate many of the same neurones. This means that a particular pattern of interneuronal communication must be maintained through the years for every memory whether important or trifling, but presumably one cell may be involved in many different memories.

9   This conception demands detailed discrimination by nerve cells so that they can discharge differently according to the pattern of afferent excitatory or inhibitory influences.

10   It is beginning to look as though this discrimination must be laid down by experience in the molecules of the nerve cell body, and the question is whether the synthesis perhaps of a particular protein in the nerve cell during consolidation of learning could determine the future reactivity of the cell in such a way as to preserve a particular threshold.

Man always applies his latest discovery to the understanding of obscure problems. Ancient man may have thought of memory as an impression like seal on wax. Tubes of vital spirits were a later conception, while in more recent times electrical currents, switching points and telephone wires dominated thought.

With computers the analogy became more complex, certainly more realistic, and the science of kybernetics has developed rapidly. Now this latest analogy is with the genetic information storage and transfer-mechanisms which have been unravelled in recent years.

As time passes and further research work develops, these ideas will all be seen in a different perspective. Young (1966) has explained the great difficulties in accepting Hydén's early ideas in 1960, but if his suggestions on modifications of RNA in relation to learning can somehow be married to the vast conceptions of the students of kybernetics, then indeed I think we may take an important step forward.

I am indebted to Mrs Freda Newcombe, D. PHIL., and Dr Robert Young for assistance in preparing this paper.

M

REFERENCES

ADAMS R.D., COLLINS and VICTOR M. (1962) In *Physiologie de l'Hippocampe*, p. 273. Series Colloques Internationaux, No. 107, Paris

BICKFORD R., MULDER D.W., DODGE H.W., SVIEN H.S. and ROME H.P. (1958) In *Brain and human behaviour*. *Res. Pub. Ass. nerv. ment. Dis.* **36,** 227

GAITO J. (1966) *Molecular Psychobiology*. C.C.Thomas, Springfield, Illinois

GERARD R.W. (1961) In *Brain Mechanisms and Learning*, pp. 29 and 183. Blackwell, Oxford

GRISSOM R.J., SNEDFELD P. and VERNON J. (1962) *Science* **138,** 429

HYDÉN H. (1966) In *Neurosciences Research Symposium*, Vol. I. M.I.T. Press, Cambridge, Mass.

KLEINSMITH L.J. and KAPLAN S. (1963) *J. exp. Psychol.* **65,** 190

MELTON A.W. (1963) Implications of short-term memory for a general theory of memory. *J. Verbal Learning and Verbal Behaviour* **2,** 1

MILNER B. and PENFIELD W. (1956) *Trans Am. Neurol. Assoc.* June 1955, **42**

MILNER P.M. (1957) *Psychol. Rev.* **64,** 242

RICHTER D. (1966) *Aspects of Learning and Memory*. Heinemann, London

RUSSELL W. RITCHIE (1959) *Brain: Memory: Learning*. Clarendon Press, Oxford

RUSSELL W.R. and NEWCOMBE F. (1966) *Aspects of Learning and Memory*. Heinemann, London

SCHMITT F.O. (1966) In *Neurosciences Research Symposium Summaries*, Vol. I. M.I.T. Press, Cambridge, Mass.

SCOVILLE W.B. and MILNER B. (1957) *J. Neurol. Neurosurg. Psychiat.* **20,** 11

TALLAND G.A. (1965) *Deranged Memory*. Academic Press, London

TRIESMAN A.M. (1964) *Am. J. Psychol.* **77,** 533

VICTOR M. and YAKOVLEV P.I. (1955) *Neurology* (Minneap.) **5,** 394

WELFORD A.T. (1958) *Ageing and Human Skill*. Oxford University Press, London

WILLIAMS M. (1950) *J. Neurol. Neurosurg. Psychiat.* **13,** 30

WILLIAMS M. and SMITH H.V. (1954) *J. Neurol. Neurosurg. Psychiat.* **17,** 173

WILLIAMS M. and PENNYBACKER J. (1954) *J. Neurol. Neurosurg. Psychiat.* **17,** 115

YOUNG J.Z. (1966) *The Memory System of the Brain*. Oxford University Press, London

CHAPTER TWELVE

PROTEIN SYNTHESIS AND THE
MOLECULAR BASIS OF MEMORY

P. N. CAMPBELL

So far three types of biological memory have been identified. The first is genetic memory and is the one most clearly understood. This has been shown to reside in the base sequence of DNA. It is now clear, therefore, that genetic memory has a structural basis which is easily comprehensible in chemical terms. The second is immunological memory, whereby the body responds to the presence of a 'foreign' substance and remembers the substance when challenged on a subsequent occasion. The exact basis of this phenomenon is not yet so well understood, but good progress is being made and it is highly probable that a structural basis will be found. The third type of memory is a function of the brain. This is, of course, the form of memory that is best known, but it is also the least understood. Because of the great inherent interest in the subject and because such excellent progress has been made in our understanding of the first two memory mechanisms, experiments concerned with the elucidation of the third type have been given great prominence and a good deal of effort is now devoted to research in this topic.

I thought that in this chapter I should consider first the theoretical basis for the experiments which have so far been conducted. I plan, then, to make some comments on the present state of our knowledge of the role of nucleic acids in protein synthesis in animal cells. Since I am a biochemist, I do not propose to consider the experimental techniques that have been used to assess the acquisition of learned behaviour.

THE THEORETICAL BASIS OF MEMORY

As explained, the only clearly understood type of biological memory involves the base sequence of DNA. For this reason hypotheses concerning memory transfer which are based on DNA naturally have a greater credence than alternative hypotheses. It seems, too, to be theoretically possible

...For ØX-174 DNA; or for about <u>3000 pages</u> like this

for E. coli DNA.

Fig. I. Imaginary base sequence of DNA from the bacteriophage Ø × 174 (from Bollum).

for DNA to have this role. I may illustrate this from a recent demonstration of Dr Bollum. In Fig. 1 is shown an imaginary base sequence of the DNA from the $\emptyset \times 174$ phage. This is the smallest, and hence the simplest, form of DNA known. By comparison the base sequence of DNA from *E. coli* which is 1.1 mm long would cover 3000 pages of an average textbook. One can imagine how many pages the base sequence of the DNA from a differentiated cell would cover. Hence one starts from the basic assumption that a structural basis for memory is possible.

There are two obvious ways in which DNA could be involved. In the first a signal would be received by the nervous system from an external media for a time sufficient for the brain to memorize it. This imprint could be effected either by causing to be made more RNA with a specific base sequence (instructive theory) or by selecting an area of the existing DNA to act as a template for RNA synthesis (selective theory).

The basic concepts for the control of protein synthesis in bacteria worked out by Jacob & Monod are shown in Fig. 2. The regulator gene is a stretch of DNA which is responsible for directing the synthesis of a repressor molecule. The substance, which is responsible in this case for inducing the bacterium to synthesize a particular enzyme, is known as the inducer. The inducer combines with the repressor and this combines with the operator which is a gene controlling the synthesis of the operon which is the stretch of DNA where is held the information necessary for the synthesis of the enzyme, or enzymes, being induced. The correct m-RNA is now synthesized which in turn leads to the synthesis of the correct polypeptide chain or chains. I would like to point out two things concerning the inducer and the repressor in this model. The model really became much more plausible after it had been shown that the combination of a variety of different substances with a protein could affect its tertiary structure. This made it more likely that the repressor would be a protein and the recent work of Gilbert and Mueller-Hill (1966) and of Ptashne (1967) has made this seem rather probable. These workers have isolated the repressors of two genes and shown them almost certainly to be proteins. Moreover, one of the isolated repressors binds specifically and with high affinity to DNA. This is the first clear-cut evidence that repressors block the transcription from DNA to RNA by binding directly to the DNA. Hence it is now perfectly feasible for us to consider a variety of substances combining with a protein repressor and so altering its shape that its ability to combine with the DNA operator is affected.

The translation of these ideas to memory models in neurones is shown in Fig. 3. Model (b) is very similar to that just described, the only difference

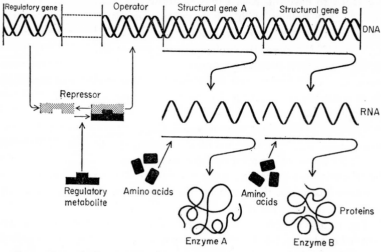

FIG. 2. Control of protein synthesis by a genetic 'repressor' was proposed by François Jacob and Jacques Monod. A regulatory gene directs the synthesis of a molecule, the repressor, that binds a metabolite acting as a regulatory signal. This binding either activates or inactivates the repressor, depending on whether the system is 'repressible' or 'inducible'. In its active state the repressor binds the genetic 'operator', thereby causing it to switch off the structural genes that direct the synthesis of the enzymes (from J.P.Changeux, *Scientific American*, April 1965, p. 37).

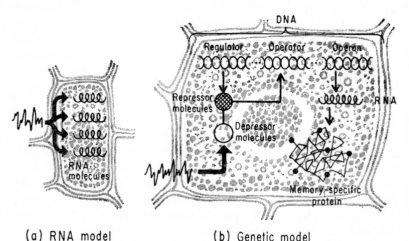

(a) RNA model   (b) Genetic model

FIG. 3. Molecular basis of memory (from *Scientific Research*, 1966, **1**, No. 6, p. 19).

being that the impinging electrical impulse causes a change in the structure of the repressor molecule. At present the mechanism directly connecting the impulses with molecular alteration is unknown as is the connection between memory-specific protein and electrical impulse response of the neurone to stimulation. This model is based on the selective theory. In contrast, model (a) is based on the elective theory. Here RNA is caused to be synthesized.

In comparing these two models let us deal with the more implausible one first, i.e. model (a). This model gained ground as a result of the work in which learned behaviour was transferred by injecting brain tissue extracts from one animal to another. It has been difficult to repeat this work (see Byrne *et al*, 1966) and I will not consider the controversial evidence here. From the practical viewpoint the experiments seem very unlikely to succeed since the chance of RNA injected either intraperitoneally or intracisternally reaching the appropriate neurone seems very remote. On theoretical grounds this elective hypothesis would be contrary to our present concepts.

The argument goes as follows. The lifetime of an RNA molecule is much shorter than the rate of decay of memory so that the new RNA must be continuously replicated as a result of the original stimulus. With the exception of RNA viruses, RNA synthesis within the cell is gene-directed, i.e. is dictated by the base sequence in the DNA molecule. Moreover, it is known that there are complex patterns of animal behaviour which are inherited and are not acquired through experience. Such behavioural patterns must derive from the DNA directed biosynthesis of RNA. There is, therefore, already a framework for a molecular mechanism mediating behaviour in the cell so it seems unlikely that a new system autonomous from the genetic system should also exist.

The elective model (b) seems much more plausible since it fits into the functional scheme of the genetic apparatus. Bonner (1966) suggested that the genes which are switched on synthesize substances (proteins) which provide for the passage of electrical impulses through specific synapses. In this case memories are stored as facilitated synaptic connections.

SOME EXPERIMENTS RELEVANT TO THE ABOVE HYPOTHESES

We may mention two sets of experiments that suggest that protein synthesis is involved in the mechanism of memory. At the University of Michigan, Agranoff and his group have worked with goldfish. These are chosen because they have relatively primitive brains, learn quickly and can retain what they are taught for a considerable time. The fish were trained to avoid an electric shock by swimming over a hurdle placed in the aquarium. Immediately before training, the experimental group were injected intracranially with the drug puromycin or cycloheximide, both drugs that interfere with protein synthesis, as described later. The control fish could remember the response for weeks, but the injected fish were unable to retain the knowledge for more than a few hours. (The drugs blocked permanent memory but did not affect short-term memory.) They got the same results when the drugs were given immediately after a learning session, but not if the injections were delayed for more than an hour after a session.

Agranoff and his colleagues have also shown that the protein synthesis is more rapid during training. They did this by following the incorporation of labelled amino acid into brain protein. The rate of synthesis of brain RNA was not affected by either the drugs or training. Thus these experiments support the concept that there is an obligatory protein synthetic step in long-term memory formation (Agranoff, Lim, Casola and Brink, 1967). Barondes and Cohen (1967) have carried out similar work with mice. The inhibitors were administered intracerebrally in the temporal region of the brain by the method of Flexner. They were given a few hours before training. By varying the times of injection and correlating these with the effect on memory at different times after training they concluded that protein synthesis is not required for learning or for memory within several hours after learning. However, protein synthesis mediated the 'long-term' phase of memory storage.

THE ROLE OF NUCLEIC ACIDS IN THE SYNTHESIS OF PROTEIN IN THE NORMAL DIFFERENTIATED CELL

In order to assess the interpretation of the above experiments it is necessary to understand the present state of our knowledge concerning the normal differentiated cell. It is not possible, of course, to cover adequately the subject in this article so I will select a few points.

Firstly we have invoked the role of a gene regulator. This concept was originated to explain enzyme induction in bacteria and, as I have explained, there is now considerable support for the theory. It is, however, a big step to say that the regulator of DNA in a differentiated cell is also a protein, and to date we have to admit that there is no evidence. For many years histones, the basic proteins associated with DNA, have been considered likely candidates as gene regulators, but they have failed to live up to expectations, so the subject is still completely open.

Secondly, if new protein is to be synthesized, then RNA must also be made. This is shown quite clearly in Fig. 2. There is conflicting evidence as to whether training in mice affects the rate of synthesis of RNA, but the balance seems to be against this. It is not easy to establish this point, for RNA is being made continuously and much of it is apparently degraded almost as quickly. Thus it is likely to be difficult to detect an increase in the synthesis of any particular RNA by the injection of precursors such as (^{3}H) uridine. The drug actinomycin has also been used to block the synthesis of m-RNA. This is a rather unsatisfactory method, for

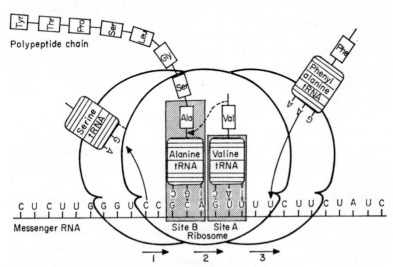

FIG. 4. Synthesis of protein molecules. The ribosome is shown to consist of two parts, the large (50S) subunit is on top and the smaller (30S) subunit is underneath. The arrows indicate that the addition of an amino acid to the growing peptide chain requires the movement of the ribosome with respect to the messenger RNA (from F.H.C.Crick in *Scientific American*, October 1966).

the drug is very toxic and the animals are sick. It seems unlikely, therefore, that it will be established in the near future whether RNA synthesis is involved.

Thirdly we should consider the use of puromycin. The present ideas on the synthesis of polypeptide chains on the ribosomes are depicted in Fig. 4. This shows that there are two sites on the ribosome for the transfer RNA amino-acid complex. The transfer RNA moves from site A to site B and concomitantly the amino acid with its transfer RNA is added to the end of the growing polypeptide chain. The structure of puromycin is very similar to that of the amino-acid transfer RNA complex; the difference being that there is no polynucleotide chain to base pair with the messenger RNA. Thus the puromycin becomes attached to the growing chain and the chain is removed from the ribosome. The puromycin does two things, therefore—it stops protein synthesis and releases a polypeptide. The effect of puromycin on memory could be due to either of these two effects. There are several theories based on the effect of peptides on memory storage (e.g. Stone, 1967). Apart from theories de Wied and his colleagues in Utrecht have obtained evidence of the effect of ACTH and similar peptides on long-term and short-term memory (de Wied and Bohus, 1966). The experiments were conducted with rats subjected to avoidance conditioning. In addition to ACTH the shorter peptide melanocyte-stimulating hormone (MSH), which has an identical sequence to part of ACTH, was also tested. MSH and ACTH had a similar effect on 'short-term' processes. In contrast, pitressin (posterior pituitary extract) had its effect on long-term memory storage.

The effect of cycloheximide was also mentioned in the last section and its use might be thought to have avoided the complications arising from the use of puromycin. That it is an effective inhibitor of protein synthesis is clear, but there is some controversy as to its mode of action (see Korner, 1968).

Finally I must mention that it has so far been extremely difficult to demonstrate conclusively the role of m-RNA in animal cells. Few would dispute the likelihood that the scheme worked out for bacteria also in general terms holds for differentiated cells. However, evidence is very fragmentary. Animal ribosomes respond to synthetic messenger RNA such as polyuridylic acid, but not even reticulocyte ribosomes clearly respond in a specific manner to natural RNA from different sources (Hunt and Wilkinson, 1967). Furthermore, the manner in which m-RNA travels from the nucleus to the cytoplasm is not established. I mention these points only to emphasize to those who are not in the field of molecular biology that our theories are often ahead of our experiments and that nowhere is this more true than in theories of memory.

REFERENCES

AGRANOFF B.W. (1967) *Scient. Am.* June, 115

AGRANOFF B.W., LIM R., CASOLA L. and BRINK J.J. (1967) 1st International Meeting of the International Society for Neurochemistry, Strasbourg

BARONDES S.H. and COHEN H.D. (1967) 1st International Meeting of the International Society for Neurochemistry, Strasbourg

BONNER J. (1966) In *Macromolecules and Behaviour.* Ed. GAITO J. Appleton-Century-Crofts, New York

BYRNE W.L., SAMUEL D., BENNETT E.L., ROSENZWEIG M.R., WASSERMAN E. *et al* (1966) *Science* **153,** 658

DE WIED D. and BOHUS B. (1966) *Nature (Lond.)* **212,** 1484

GILBERT W. and MUELLER-HILL B. (1966) *Proc. natn Acad. Sci. U.S.A.* **56,** 1891

HUNT J.A. and WILKINSON B.R. (1967) *Biochemistry* **6,** 1688

KORNER A. (1968) In *Interaction of Drugs and the Subcellular Components of Animal Cells.* Ed. CAMPBELL P.N. J.&A.Churchill Ltd, London

PTASHNE M. (1967) *Nature (Lond.)* **214,** 232

STONE A.B. (1966) *Scient. Res.* Nov. 6.

CHAPTER THIRTEEN

THE USE OF IONIZING RADIATION IN NEUROLOGY

A.M.JELLIFFE

INTRODUCTION

In this chapter, there is no space to mention any of the diagnostic applications of ionizing irradiations that are obviously so important in neurology, neurosurgery and radiotherapy today, but only to consider the effects of ionizing irradiation on normal nervous tissue and the advantages and disadvantages that ionizing irradiation offers in the management of some diseases of the nervous system.

Whether beneficial or harmful, the final biological effect of all ionizing irradiations on living tissues depends upon the production of chemical changes in the cell. Some types of cell are more sensitive than others to these chemical changes and the chromosome is the most sensitive part of the individual cell.

Except with the heavy particle forms of ionizing irradiation, the intensity of the changes produced in the living cell depend partially upon the presence of an adequate supply of oxygen. Cellular radiosensitivity can be greatly diminished by anaemia, or by reducing the blood supply to the tissue concerned. This point will be referred to again when discussing the place of radiotherapy in the treatment of cerebral tumours.

Various types of ionizing irradiation with different characteristics have been used in the treatment of diseases of the nervous system. It may be useful to mention briefly these characteristics as obviously they affect their clinical usefulness.

EXTERNAL IRRADIATION

Until recently, external irradiation of the nervous system has been carried out with X-rays generated at 180–250 kV. Over the last few years increasingly powerful machines have become more widely available which emit either more penetrating X-rays generated at 1 or more million volts, or gamma-rays from large radiocobalt sources, which are equivalent to X-rays generated at more than 2 million volts. These more powerful megavoltage machines are extremely expensive, but their use is justified for three reasons. Firstly, a smaller dose of irradiation is absorbed in the bone, so there is less risk of late bone necrosis. Secondly, megavoltage rays are more penetrating, and about twice as much of each incident

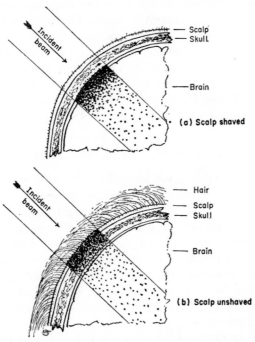

FIG. 1. Shows the skin and hair follicle sparing effect of megavoltage irradiation, when the hair is shaved (a); the maximum dose is brought up to the hair follicle level by a thick layer of unshaved hair (b).

beam reaches the centre of the skull. It is therefore easier to deliver an adequate dose accurately to a deep-seated tumour, with less irradiation of neighbouring normal structures. Lastly, with megavoltage irradiation the maximum dose is beneath the skin. The patients have no skin reaction, the hair and follicles are spared. From the patient's point of view, this is a very considerable improvement. But in order to make the best use of this effect it is usually necessary to shave the head. For example, using 4 MeV X-rays the maximum dose is just over 1 cm deep to the surface (Fig. 1).

If the patient has a nice thick layer of hair at the time of irradiation, this will bring the maximum dose up to the surface of the skin and produce the most effective hair follicle damage.

External irradiation may also be delivered with beams of particulate irradiation, such as particles, protons and electrons. Whereas X-rays and gamma-rays are attenuated expotentially in their pathway through the skull and there is always an exit dose, beams of heavy particle irradiation have a definite finite range so that when that range has been reached no irradiation continues on through the tissues. There is also a final build-up in energy absorption at the end of the beam, so that a high dose at depths can be achieved with a single beam. But this type of irradiation requires extremely expensive and complicated equipment. For example, most of the work on pituitary destruction has been carried out by Tobias and his associates (1964) using the Berkeley 184 inch synchrocyclotron and very few machines capable of producing beams of this type exist.

INTERNAL IRRADIATION

Irradiation is very rarely given internally, although in the past attempts have been made to treat brain tumours by the insertion of radioactive material. Nowadays, the only intracranial region which is commonly implanted is the pituitary gland in an attempt to produce its complete destruction, particularly in breast cancer. The first workers implanted radon sources emitting rays, but these rays produced a dangerously high dose in structures immediately adjacent to the pituitary gland. Sources emitting β-particles are now used and the very limited range of these particles allows intense pituitary irradiation, minimizing the effects on neighbouring structures.

Intrathecal instillation of radioactive colloidal solutions has been suggested in the treatment of patients with tumours such as the medulloblastoma, which may spread through the CSF space (Lewis, 1953; Kerr, Schwartz and Seaman, 1954; Rubenfield and Zeitel, 1956). But many of the tumour deposits along the spinal cord and nerves are bulky, and irradiation by this method will give adequate treatment only to the surface of these deposits. Other ingenious methods of irradiating the brain internally have been devised. For example, stable boron is concentrated in some brain tumours. After its administration, the tumour is irradiated with neutron beams. Not only do these neutrons have a direct ionizing effect on the tumour, but the boron becomes radioactive. So far such methods have not proved to be of much practical value.

THE EFFECTS OF IONIZING IRRADIATION ON THE NORMAL CENTRAL NERVOUS SYSTEM

The treatment of the central nervous system with ionizing irradiation is potentially dangerous due to its many effects on normal tissues.

INDIRECT EFFECTS

Irradiation of any part of the body may give rise to 'radiation sickness'—a syndrome characterized by loss of appetite, nausea, vomiting, somnolence and depression. The cause of the syndrome is unknown, but the symptoms can be very similar to those produced by primary intracranial disturbances, and its differentiation of obvious importance.

DIRECT EFFECTS ON THE CENTRAL NERVOUS SYSTEM

Until recently it was generally accepted that the central nervous system was extremely resistant to radiation effects. Only 22 years ago, one group of radiotherapists voiced the opinion that there was no clinical evidence to suggest that there would be any danger in increasing tumour doses to 10,000 or even 15,000 r (Pierce, Cone, Elvidge and Tye, 1945), and in 1941 the largest tumour dose in the literature—a total of 25,000 r—was reported (Kaplan, 1941).

Over the last 20 years an enormous amount of work on the subject has led to a completely different view. It is now generally accepted that radiation damage to the nervous system can be produced not only by its effect on the vascular supply, but also because of direct effects upon the specialized cells of the nervous system.

Acute effect

Small doses of radiation to the whole brain can produce EEG changes (Sams, Aird, Adams and Ellimaus, 1964), alteration in the behaviour of experimental animals and direct or indirect stimulation of the autonomic nervous system (Zeleny, 1956; Arbit, 1964; Kinneldorf and Hunt, 1964). Experimentally, *large doses* of radiation can produce an immediate response which may prove fatal within 3 days. Under these circumstances, death has been attributed to intense oedema of the meninges and brain leading to progressive symptoms cumulating in a pressure cone (Ross, Leavitt, Holst and Clemente, 1954). The normal blood-brain barrier is damaged, affecting its permeability to various substances (Clemente and Holst, 1954). Such massive effects are not seen in the treatment of patients, but there is no doubt that the irradiation is sometimes associated with a temporary increase in symptoms of brain tumours which can often be alleviated by giving the patient corticosteroids, diuretics or intravenous hypertonic solutions. Although it has never been proven, this increase in symptoms after irradiation has always been attributed to oedema of the brain, and because of this theoretical risk, large single doses to the brain are avoided by many radiotherapists.

Some very interesting work has been published recently which casts doubt on this widely accepted hypothesis. Fine vinyl catheters were inserted permanently into the cervical subarachnoid space of dogs. There were no complications and the catheters recorded accurately all normal variations in the cerebrospinal fluid pressure. Five dogs were kept as controls and 5 dogs were given 1000 r, 2000 r and 4000 r whole brain irradiation in one single dose. Spinal fluid pressure changes were then recorded for 14 days.

Very few clinical effects were noticed in animals given 1000 r or 2000 r but all animals given 4000 r became lethargic, ataxic and weak, and by the fourteenth day all were very ill. At autopsy no evidence of pressure cone formation or flattening of sulci was found. None of the animals showed any rise in cerebrospinal fluid pressure throughout the entire experiment (Redmund, Rinderknecht and Hudgins, 1967). If this work is confirmed, another explanation will be necessary for the post-radiation brain syndrome.

Chronic effects

Experimentally, animals surviving the acute effects of whole brain irradiation pass through a latent period of many months after which necrosis supervenes. Similar delayed effects are possible after the clinical use of ionizing radiation. Although some of these chronic changes are secondary to vascular occlusion, there is also a direct effect which is 'strikingly effective for white matter' (Arnold, Bailey and Harvey, 1954). Starting as a demyelinization there is progression to necrosis of all constituents of the white matter; necrosis of this type has been recorded in the absence of serious vascular changes (Wachowski & Chenault, 1945). Obviously the site of the damaged nervous tissue affects the clinical picture produced; involvement of the motor areas, brain stem or spinal cord may produce serious deficits leading to complete paralysis and death.

Some knowledge of the tolerance of the normal central nervous system to radiation is essential if the radiotherapist is to avoid or minimize the production of serious complications, but such knowledge has been difficult to acquire. Obviously when patients are deliberately given large doses of irradiation to the brain and spinal cord, they are treated in this way because a pathological process is already present. The development of complications at a later date may be due to either a recurrence of this original pathology, or to radiation damage in the original pathological process, or to radiation damage in the adjacent normal nervous tissue or to a combination of all three processes.

RADIATION TOLERANCE OF THE BRAIN
Lindgren (1958) has published an excellent review of the subject to which he has added his own original observations. In the literature, he found

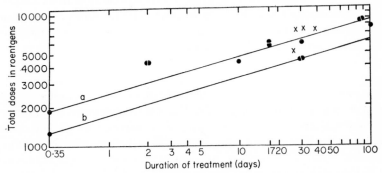

FIG. 2. Graph published by Lindgren, illustrating the relationship between dose and time, in seventeen cases of post-radiation necrosis of the brain.

only fifteen case reports which included both the exact details of irradiation delivered and the post-mortem findings which provided histological verification of the diagnosis of brain necrosis (Scholtz and Hsu, 1938; Wachowski and Chanault, 1945; Pennybacker and Russell, 1948; Zeman, 1949; Boden, 1950; Foltze, Holycke and Heyl, 1953; Arnold, Bailey and Harvey, 1954; Malamud, Boldrey, Welch and Fadell, 1954; Dugger, Stratford and Bouchard, 1954). Two patients had been treated with supervoltage irradiation and he discarded these because of the lack of a reliable factor for conversion of the biological effects of supervoltage irradiation to those occurring in the 200 kV range. To the thirteen remaining, he added four new cases—a total of seventeen (Fig. 2). Treatment times varied from one single treatment, up to a maximum of 105 days. Lindgren drew two lines on his graph. The upper line follows an analysis of the material according to the smallest square root method—

N

this line smoothes out the points and gives an average value of the time-dose effects which produced necrosis. The lower line represents the lowest time-dose level which produced cerebral necrosis in Lindgren's series.

It is important to remember that even if a human brain is irradiated to doses indicated by the top line, necrosis will not necessarily develop. Many brains have been irradiated to dose levels greatly exceeding those indicated by this line and the patients have survived without clinical evidence of necrosis. And it may be necessary to irradiate to a higher dose if the aim of treatment is the destruction of a tumour which would otherwise kill the patient. Under these circumstances a calculated risk must be taken. But if the aim of the radiotherapist is to treat a tumour arising from structures adjacent to the central nervous system, then every possible precaution must be taken to ensure that the limits of tolerance are not exceeded. The dose levels that are safe are approximately 3500 r in 3 weeks, 4000 r in 4 weeks or 5000 in 6 weeks.

RADIATION TOLERANCE OF THE SPINAL CORD

A reasonable estimate of the radiation tolerance of the human spinal cord may be arrived at in a similar way. The first comprehensive publication on the subject was by Boden (1948) who described six patients with radiation myelitis of the cord; 2 years later he reported three more (Boden, 1950). Other reports have appeared on the subject (Smithers, Clarkson and Strong, 1943; Stevenson and Eckhardt, 1945; Greenfield and Stark, 1948; Dynes and Smedal, 1960; Vaeth, 1965), and the findings of some of these reports are summarized in Fig. 3. The tolerance of the spinal cord appears to be very similar to that of the brain. Again, it must be emphasized that damage to the cord will not necessarily be caused by doses of this order. For example, Boden (1950) indicated that of forty-one patients given 4300 r or more in 17 days, eleven, or 25 per cent, developed manifestation of spinal cord damage. If the spinal cord has to be included in the treated volume in order to cure a malignant tumour, a calculated risk is taken. Unnecessary irradiation of the cord must be avoided or minimized.

When irradiating the central nervous system another factor that should be considered is the volume included. Boden suggested that the maximum tolerated dose to the spinal cord and brain stem should be either 4500 r in 17 days with small volumes and 3500 r in 17 days for large volumes.

The symptoms of spinal cord damage appear one to many months after irradiation; the greater the dose the more rapid and severe the damage. Not all patients develop irreversible changes. Boden described a transient stage of numbness and hyperaesthesia of either the upper or lower extremities, or both, often radiating down from the neck. Attacks were often precipitated by neck movements and frequently the patient had a stiff neck. There were no objective signs of nervous system damage and the symptoms settled within a year. These findings are thought to be due to mild damage to the ascending pathways in the posterior and lateral columns.

Permanent changes may be preceded by the transient symptoms recorded above. Muscle weakness and paralysis may appear and then stop

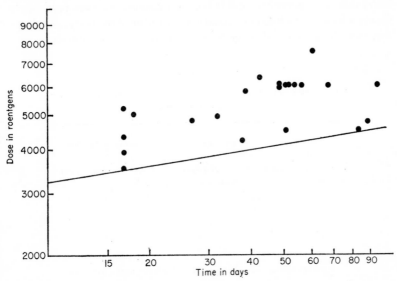

FIG. 3. Graph plotting dose against time, with twenty-two cases of post-radiation myelitis.

at a certain stage; or the weakness may progress slowly, leading to partial or complete transection of the cord. The cerebrospinal fluid pressure is normal, myelography shows no obstruction and there is no evidence of raised intracranial pressure. The cerebrospinal fluid is normal except for possibly a slight rise in protein. Histological examination of the cord shows either necrosis and demyelinization of the cord or ischaemic changes secondary to vascular thickening, or both.

PERIPHERAL NERVES
Peripheral nerves are, on the whole, resistant to irradiation. But very high doses may be followed in a few years by dense connective tissue fibrosis

which can produce severe damage in peripheral nerves entrapped in a constricting mass of scar tissue. This has been seen most often in the past as damage to the brachial plexus, but recently we have seen damage to the femoral nerves and nerves in other areas, produced by supervoltage irradiation. For example, when treating carcinoma of the bladder, an area of subcutaneous fibrosis can be produced by supervoltage irradiation on the buttock, without any trouble to the patient. But a similar area over the femoral triangle can entrap the femoral nerve with serious results. This complication is now recognized and can be prevented by avoiding heavy doses of supervoltage irradiation over important nerves, particularly nerves which are enclosed in a small compartment of the body, and therefore specially prone to damage by a constricting band of fibrosis.

TABLE I

Intracranial neoplasms in children, long-term survival after radiotherapy (Bouchard)

Observation period (years)	Survival rate		Quality of survival		
	Total patients	Survivors	Good	Partial	Poor
5	119	50 (42%)	39	8	3
10	86	31 (36%)	25	5	1
15	45	14 (31%)	10	3	1
20	14	6 (43%)	5	1	—

Post-operative radiotherapy: 78.   Radiotherapy alone: 41.

IRRADIATION OF CHILDREN

Although it is probable that the central nervous system continues to develop until the age of about 15 years, there is no evidence that radiotherapy as used in the treatment of brain tumours leads to retardation of the mental development and somatic growth (Table 1). And certainly there is no evidence whatsoever that the child's pituitary gland is damaged by this treatment. Some authors advise a small reduction in the generally recommended dose when treating children. For example, Bouchard (1966) advises a 10 per cent reduction for all children and Richmond (1953) recommends 50 per cent of the adult dose under 1 year, 75 per cent between 1 and 5 years and the full dose from 8 years upwards.

While agreeing in principle with the need to prevent possible damage in later life, the author finds this approach difficult to understand. When dealing with a lethal disease, some risks must be taken, and the evidence so far is that radiotherapy as used in the treatment of brain tumours does not reduce the child's chances of survival into normal adult life. Most reports of brain tumours treated in childhood indicate that older children do better than young children. This certainly reflects the more rapid growth of cerebral tumours in the younger group, but it possibly also indicates a natural and praiseworthy desire to avoid the risk of producing late damage in these very young children—a desire which appears to be stimulated more by emotion than by factual knowledge. It is interesting to see that the Manchester school of radiotherapy (Paterson, 1963) makes no special reference at all to the need for a smaller dose of irradiation when treating children.

The long-term results of treating brain tumours in children can be extremely good—much better than the results with most other tumours of childhood. And, as already indicated, there is no evidence that mental retardation follows this treatment. But obviously, unnecessary irradiation to the brain must be avoided and the brain and spinal cord in the child must be protected with great care when irradiating neighbouring structures.

RADIOTHERAPY OF BRAIN TUMOURS

In clinical practice, ionizing irradiations are used most commonly in the management of diseases of the nervous system in the treatment of tumours. There is no doubt that they can sometimes be extremely useful; on other occasions they are of doubtful value and sometimes their use appears to be harmful to the patient. If they are to be used to their best advantage, each case must be assessed individually and the possibilities of surgery, radiotherapy or masterful inactivity carefully considered.

There are certain general principles which are worth remembering when considering the possible value of radiotherapy in brain tumours. Radiotherapy is most likely to produce a good result when the tumour is radiosensitive, when it is of small volume, when its blood supply is adequate and has not been interfered with by surgery or previous radiotherapy, and when it is situated in a relatively unimportant part of the brain. The patient is more likely to appreciate the result if the tumour originally produced unpleasant symptoms which then disappear.

Obviously some of these points may make a tumour even more

curable by surgery, and there should be no competition between the two methods which should be regarded as complementary. But the closest co-operation between physician, surgeon and radiotherapist is needed to allow the patient the maximum benefit of all methods of treatment. For example, radiotherapy could perhaps be considered earlier in the management of some radiosensitive tumours which are at present usually irradiated only after incomplete surgical removal, which sometimes carries a high mortality rate. Some patients might be better off if radiotherapy was withheld, after surgical removal of a low-grade, relatively radioresistant tumour. In this situation there is often no real need for immediate radiotherapy, and if post-operative radiotherapy is used, it is almost impossible to know if the patient has benefited or not. Unnecessary radiotherapy can only do harm, and it is better to wait until a real need for it arises.

GLIOMAS

The potential value of radiotherapy in the management of the gliomas can be assessed more easily if these tumours are collected into a small number of main groups.

Glioblastoma multiforme (Kenohan, types III and IV)
If untreated, these quickly growing, locally invasive tumours are rapidly lethal. About 90 per cent of patients are dead within 12 months (Elvidge, Penfield and Cone, 1935; Bailey, 1948; Frankel and German, 1958; MacCarty, 1962; Taveras, Thompson and Pool, 1962). When the prognosis of a tumour is so appalling, it should be easy to see if any form of treatment has a beneficial effect. Some authors believe that radiotherapy does prolong life. Roth and Elvidge (1960) analysed more than 300 cases and found that after surgery alone the average survival was 9.8 months; after surgery and radiotherapy it rose to 17.4 months. Taveras *et al* (1962) reported 425 cases and found that after radiotherapy in addition to surgery, a few more patients survived for a longer time. In both those studies, patients not surviving for 1 month after surgery were excluded in an attempt to make a fair comparison between those treated with surgery alone and those treated by surgery and radiotherapy. But even this adjustment cannot completely exclude some bias in favour of those patients treated with radiotherapy. Bouchard (1966) believes that worthwhile life can be prolonged by radiotherapy: of a total of 171 patients treated by surgery and radiotherapy, 44 per cent survived 1 year and 7 per cent survived 5 years. Of the thirteen 5-year survivors, ten were in excellent condition and three were partially disabled.

We must conclude that, as yet, there is no convincing evidence that radiotherapy prolongs the life of patients with glioblastomas. But there is no doubt that radiotherapy can produce symptomatic improvement, and this improvement may continue for a long time. Because of this unpredictable response and because cure is almost impossible, a decision regarding the advisability of radical treatment should be reached only after very careful consideration of all the facts. For example, it would appear morally wrong to give a prolonged course of radiotherapy for a glioblastoma which has been discovered in an old man who is alone in the world and free of responsibilities, who has longstanding severe neurological deficit which is probably irreversible and in whom the tumour has fortunately produced a complete lack of appreciation of his serious condition. On the other hand, radiotherapy must always be considered if the victim is a young man with recent neurological changes, a family to support and an acute awareness of the nature of his condition. To such a patient and his family, even a short period of benefit may be of incalculable value, but on the whole these tumours should be regarded as relatively radioresistant: a reasonable tumour dose would be about 6000 r in 7 weeks.

Astrocytoma (Kenohan, types I and II)
These tumours can be considered more easily if they are divided into supratentorial and subtentorial groups.

Supratentorial astrocytoma. On the whole these are slowly growing, diffusely infiltrating tumours, subject to periods of apparent regression and sudden exacerabation, the latter presumably due to small haemorrhages into cystic spaces. The value of any treatment is very difficult to assess, because of the insidious progress and considerable spontaneous variation. But it seems probable that complete surgical removal is possible in about 1 in 3 patients. One difficulty encountered by the surgeon is that the growing edge often cannot be defined. The radiotherapist is also hampered by this lack of definition and even more so by the very large size reached by some of these supratentorial tumours, which reduces their radiosensitivity.

There is some evidence that radiotherapy helps to control these tumours. In 1956, Levy and Elvidge analysed eighty-nine supratentorial astrocytomas treated in Montreal: about a half had surgery alone and the other half had radiotherapy as well as surgery. They concluded that radiotherapy was of some benefit. In 1966 Bouchard analysed 105 patients from the same centre who had been treated with surgery and radiotherapy; his series included forty-five already reported by Levy and Elvidge. The survival rate in this group was considerably better, about half the patients surviving 5 years (see Table 2).

A decision regarding the advisability of radiotherapy must always take into account all the factors referred to in the treatment of glioblastomas, but the time scale is very difficult. Treatment is more worth while even in the presence of advanced neurological complications, as very long term remissions may follow.

Infratentorial astrocytoma. These present usually as cerebellar tumours in children and young adults. Generally, they are more cystic, less invasive and have a well-defined growing edge; the rapid progression of symptoms

TABLE 2

Astrocytoma: comparison of published series. Treatment by surgery alone, or by surgery and radiotherapy

	Levy and Elvidge (1956)		Bouchard (1966)
	Surgery alone (%)	Surgery + radiotherapy (%)	Surgery + radiotherapy (%)
All cases	42–100	45–100	105–100
Survival at 1 year	35– 81	37– 82	90– 86
Survival at 3 years	22– 52	28– 62	67– 64
Survival at 5 years	11– 26	16– 36	51– 49

results in earlier diagnosis. The problems of management are therefore very different from those presented by supratentorial astrocytomas.

Levy and Elvidge reported forty-one cases, thirty-four of which were adequately followed. Twenty-nine of these had surgery alone—and twenty-two, or 76 per cent survived over 5 years. It is interesting that of these twenty-two 5-year survivors, four had tumours which were apparently removed incompletely. Bouchard (1966) reported fourteen patients treated by surgery and radiotherapy, of whom twelve were still alive after periods of 5–22 years. All were mentally normal and only one was partly disabled by an unsteady gait.

Because of their smaller size, cerebellar astrocytomas should respond better to radiotherapy than cerebral astrocytomas. Nevertheless, the results of surgery are extremely good, and at the present time there does not seem to be any evidence that radiotherapy is indicated, if these well-localized tumours are apparently excised completely. Post-operative radiotherapy may be of value when the tumour is not totally removed.

Oligodendroglioma
These uncommon tumours are usually relatively benign, being well circumscribed and slowly growing: scattered calcification is frequently seen on X-ray examination. A small number, showing more malignant microscopical and clinical characteristics, are described as oligodendro-blastoma. There are few useful reports referring to the treatment of these tumours, but recently an interesting paper has been published (Sheline,

TABLE 3

Oligodendroglioma. Results of treatment of thirty-seven cases. Five excluded because of death within 14 days of operation (Sheline *et al*, 1964)

	Surgery alone	Surgery + radiotherapy
All cases	16	16
All cases followed for 5 years	13	13
Alive at 5 years	4	11
5-year survival rate	31%	85%
All cases followed for 10 years	8	11
Alive at 10 years	2	6
10-year survival rate	25%	55%

Boldery, Karlsberg and Phillips, 1964) which refers to thirty-seven cases, none of which were examples of the more malignant oligodendro-blastoma. Five patients died within 14 days of operation. Of the remaining thirty-two, sixteen patients received post-operative radiotherapy, thus creating two approximately comparable groups. One patient in each group was lost to follow-up and two in each group received treatment less than 5 years ago. A comparison of the thirteen remaining patients in each group shows a survival rate at 5 years and 10 years, that is more than twice as good in those patients treated by radiotherapy as well as surgery (Table 3). Radiotherapy therefore appears to be of value in the management of these tumours.

Ependymoma

Most of these tumours grow slowly and are relatively benign: not more than 20 per cent grouped as ependymoblastomas are more active. It is generally accepted that the complete surgical removal of an ependymal tumour is very difficult (Bouchard, 1966; Kricheff, Becker, Schneck and

TABLE 4

Supratentorial ependymomas (5-year survival)

Treatment	No. of patients	5-year survivals	Survival (%)
Surgical removal (Ringertz and Raymond 1949)	13	2	15
Surgical removal and radiotherapy (Kricheff *et al* 1964)	9	2	22
Surgical removal and radiotherapy (Bouchard 1966)	17	11	65

TABLE 5

Infratentorial ependymomas (5-year survival)

Treatment	No. of patients	5-year survivals	Survival (%)
Surgical removal (Ringertz and Raymond 1949)	15	5	35
Surgical removal and radiotherapy (Kricheff *et al* 1964)	33	15	45
Surgical removal and radiotherapy (Bouchard 1966)	3	3	100

Taueras, 1964). The value of radiotherapy is indicated by comparing some of the published series.

Supratentorial ependymomas treated by surgery alone have a 5-year survival rate of under 20 per cent: the addition of radiotherapy increases this rate considerably. Bouchard refers to a 5-year survival rate of 65 per cent and comments on the excellent physical state of the long-term survivors (Table 4).

Infratentorial ependymomas carry in general a better prognosis, presumably because in this region small tumours produce symptoms more rapidly, allowing earlier diagnosis and treatment (Table 5). There is no doubt that many of these tumours are extremely radiosensitive. One reported patient lived for 24 years after a tumour dose of only 900 r and died of local recurrence (Kricheff *et al*, 1964). Post-operative radiotherapy is obviously essential. Perhaps the importance of radiotherapy in the management of ependymomas should be emphasized even more. It is generally accepted that most of these tumours are not curable by radiotherapy alone. Their extreme radiosensitivity suggests that results might be improved if surgery was limited to a decompression and to obtain adequate material for diagnostic purposes, and more reliance placed on energetic radiotherapy.

Prophylactic irradiation of the entire spinal axis has been recommended because of the reported tendency of ependymal cell tumours to spread along the cerebrospinal spaces (Dyke and Davidoff, 1942). One paper describing the findings at nineteen post-mortem examinations reported that six cases showed spinal cord deposits—but from these deposits the patients had had no clinical manifestation during life (Svien, Gates and Kernohan, 1949). Kricheff *et al* (1964) treated four of their sixty-five cases with prophylactic irradiation. Only one patient in the whole group developed spinal metastases and this was one of the four given prophylactic irradiation previously. Most authors now agree that prophylactic irradiation of the entire cerebrospinal space is not indicated in the management of ependymal gliomas, although some recommend this method of treatment if the primary lesion is infratentorial and projects into the cisterna magna (Moss, 1965).

Medulloblastoma
This tumour usually originates from the midline of the cerebellum and is seen most commonly in children. Attempts to cure these tumours by surgery alone are futile (Cutler, Sosman and Vaughan, 1936; Smith, Lamp and Kahn, 1961; Berger and Elvidge, 1961). Surgery is not only ineffective but it also carries a high operative mortality rate. And although there is no doubt that dissemination of tumour cells along the subarachnoid space can occur without vigorous surgical interference, the possibility that tumour dissemination is encouraged by surgery cannot be disregarded.

The medulloblastoma is the most radiosensitive tumour arising in the central nervous system and post-operative radiotherapy to the primary site and to the entire subarachnoid space is essential if cure is to be attempted.

If these facts are accepted, it is clear that radiotherapy is being sadly misused in the treatment of medulloblastoma. In this condition, energetic surgery which cannot cure and which may kill the patient has no place whatsoever. Surgery provides the only method of confirming the diagnosis histologically and it is suggested that the least traumatic way of obtaining histological proof of a cerebellar tumour in a child is by aspiration biopsy (Bouchard, 1966). If microscopical examination demonstrates an astrocytoma, surgery, probably alone, is the treatment of choice. But if a medulloblastoma is found, then without any further delay the primary site and entire subarachnoid space should be irradiated. The Manchester method of irradiating the posterior aspect of the subarachnoid space (Paterson, 1963) is a valuable technique.

TABLE 6

Cerebellar medulloblastoma in children (Bouchard, 1966)

Treatment method	Proportion of survivors			
	1 year	*3 years*	*5 years*	*10 years*
Surgery and irradiation	17/27	9/27	6/27	3/23
Aspiration biopsy and irradiation	8/10	4/10	4/10	2/7

The figures published by Bouchard confirm that this approach is the logical one. The number of patients in his series treated by aspiration biopsy and irradiation alone is small, but twice as many patients survived for long periods after treatment by this method than after more energetic surgery followed by radiotherapy (Table 6).

Bouchard (1966) recommends a tumour dose of 5000 r to the intracranial contents and 2000 r to the entire spinal canal over a period of about 9 weeks.

Midbrain and brain stem

The first problem with tumours arising in these sites is that of verifying the diagnosis by biopsy. Obviously, complete surgical extirpation is impossible and even taking an adequate biopsy may be hazardous, but surgery limited to decompressive shunts can produce remarkable temporary improvement.

From the radiotherapist's point of view gliomata arising in this region should be expected to respond better to irradiation that do the extensive tumours so often encountered in the cerebrum. Very small tumours in the brain stem and midbrain produce obvious symptoms, encouraging early investigation. Unfortunately, so often after the diagnosis of 'deep-seated mischief' in these inaccessible regions has been made, the lack of confirmatory pathological evidence encourages delay before the patient is referred for radiotherapy. This is quite illogical. Even when there is no pathological confirmatory evidence, radiotherapy must be considered as soon as the clinical diagnosis of an expanding lesion in this region has been established. Only a very small number of conditions simulating tumours will be necessarily treated and there is no evidence that irradiation of these other rare conditions is harmful. Indeed, in the past, radiotherapy has been recommended for the treatment of some, such as syringobulbia.

Delay allows the further irreversible destruction of nervous tissue, and any increase in tumour size reduces its radiosensitivity. There is therefore nothing to lose and much to gain from the early use of radiotherapy. Bouchard (1966) reports the results of treating 126 tumours in these sites. Prolonged survivals of worth-while life can be very considerable and long-term survivors usually remain in good condition throughout their survival period. Tumour doses of between 5000 and 6000 r in 7–8 weeks are usually advisable. Because of the more serious sequelae of over-treating this region, the pons and medulla should be taken to the lower of these dose levels, whereas it is reasonable to treat the midbrain to the higher dose level.

NON–GLIOMATOUS PRIMARY INTRACRANIAL TUMOURS
Non-gliomatous tumours may arise from many intracranial structures, but only some of the more common groups will be mentioned.

Meningioma
In general, these tumours may be considered to be well localized, non-invasive, slowly growing, microscopically benign and relatively radio-resistant.

When accessible, surgical removal is extremely successful and excellent long-term results can be expected. Under these circumstances, radio-therapy cannot benefit the patient, and its only effects are likely to be harmful. There is an extremely small group of these tumours which may be improved by radiotherapy and because such a very small number of patients is involved, a reliable assessment of the value of radiotherapy is almost impossible.

Radiotherapy may be considered after the removal of a meningioma under the following circumstances. Firstly, microscopical examination of the meningioma may indicate an invasive sarcoma. These appearances suggest that a rapid recurrence is likely, and that the tumour may be more radio-sensitive than the usual meningioma. Unfortunately, there is tremendous variation in the histological appearance of meningeal tumours and the microscopical findings are not related directly to their radiosensitivity. Secondly, the operative findings may suggest the possibility that the tumour has spread beyond the limits of the tissues removed at operation.

Under both these circumstances the use of radiotherapy may be considered. But many of these patients enjoy normal lives without exposure to ionizing irradiation. The variable behaviour of meningiomas makes such treatment of doubtful value, and unnecessary radiotherapy is always potentially harmful. It is suggested that a wait-and-see policy might be preferable for problems of this type and they should be carefully reassessed at frequent intervals. If a recurrence is discovered, then the use of radiotherapy should be considered. But even in this situation surgical removal of the recurrence is preferable, if it is possible.

When the meningioma is quite inoperable, the radiotherapy is often worth while. Although cure is unlikely, some tumour regression is possible with relief of symptoms. But a large tumour dose is required and the risks to normal brain tissues must be justified by the severity of the symptoms.

Vascular tumours
Cerebral angiomas are congenital malformations, often associated with congenital abnormalities elsewhere. They are not true tumours. Microscopical examination demonstrates ramifying arteries and veins which are histologically mature; most mature cells respond relatively poorly to irradiation and any reduction in vascularity of these tumours depends upon the production of a non-specific obliterative change in the abnormal vessels. Changes of this type may also be produced in normal blood vessels, supplying adjacent normal brain tissue. It is suggested that radiotherapy is indicated when an angioma is producing symptoms which are incapacitating the patient or when there is evidence of its progression over a period of time. It must be remembered that a number of these lesions can be completely excised.

Haemangioblastoma, including haemangioendothelioma, arise almost exclusively in the cerebellum: complete surgical removal is usually possible and radiotherapy is not necessary. When complete removal is impossible or when microscopical examination shows that the tumour is malignant, radiotherapy should be considered. However, these tumours are uncommon and there is no proof that immediate post-operative radiotherapy is preferable to radiotherapy as soon as a recurrence is detected. If frequent follow-up examinations can be arranged, a 'wait-and-see' policy may be reasonable, otherwise immediate post-operative radiotherapy is indicated.

Pinealomas
Most of the tumours in the pineal region are not true pinealomas and the value of radiotherapy in the management of tumours in this site is impossible to assess. For obvious reasons microscopical confirmation of a pinealoma is difficult and without this other space-occupying lesions, including ependymomas and choroid plexus cysts, may be indistinguishable from pinealomas. Very rarely true pinealomas metastasize through the cerebrospinal fluid.

Consideration of all these possibilities indicates the place of radiotherapy in the management of so-called pinealomas. A surgical shunt is essential and this is usually followed by a dramatic improvement. Post-operative radiotherapy is indicated, firstly because the relatively small size of these tumours and their precise localization allow accurate treatment without serious risk to the adjacent normal brain and secondly because some of these lesions are either true pinealomas or ependymomas and therefore potentially radiocurable. There is no indication for routine irradiation of the entire cerebrospinal fluid space.

THE PITUITARY GLAND

The normal adult pituitary gland is radioresistant. External irradiation of the pituitary with X-rays in doses of up to 10,000 r has been reported (Kelly *et al*, 1951) without any evidence of pituitary dysfunction, and damage to neighbouring structures has prevented the use of higher doses of X-rays.

Complete destruction of the normal pituitary by irradiation requires either the insertion of high-activity β-ray emitting sources or carefully collimated beams of deuterons or protons giving a dose to the whole gland of many thousands of rads (Tobias *et al*, 1964). Apart from the destructive effect of huge doses, there is no evidence that irradiation has any effect upon the normal pituitary. There is no justification for the use of small, so-called stimulating doses for the treatment of thyroid disorders, sterility or dysmenorrhea as has been recommended in the past.

TUMOURS OF THE PITUITARY GLAND

Treatment of pituitary gland tumours was first attempted in 1907 (Gramegna, 1909; Beclere, 1909) and one of these early patients, a girl aged 16 with gigantism, lost her headaches and recovered her vision; 13 years later she was in good health. Considering the primitive nature of Beclere's equipment, this was a remarkable result stimulating further trials of radiotherapy in the treatment of pituitary tumours. In 1925, Dott, Bailey and Cushing advised a trial of radiotherapy in all cases of pituitary when the vision was not threatened.

TABLE 7

Chromophobe adenoma (Henderson on Cushing's series)

Treatment	No. of cases	Recurrence free at 5 years	5-year (%)
Transsphenoidal operation	67	22	32.8
Transfrontal operation	40	23	57.5
Transsphenoidal operation and X-rays	49	32	65.3
Transfrontal operation and X-rays	31	27	87.1

The value of radiotherapy in the management of chromophobe adenomas was clearly seen in 1939 when Henderson analysed Cushing's series of 338 patients (Table 7). This analysis showed beyond doubt that the percentage of patients without recurrence at 5 years was much greater when patients were treated by X-ray therapy as well as surgery. It is not surprising that pituitary tumours are so much easier to treat with radiotherapy than the gliomas. Thorough investigation, with or without surgical exposure, usually indicates precisely the limits of the tumour and the whole volume can be encompassed precisely by carefully directed X-ray beams.

Chromophobe adenomas

The value of radiation in chromophobe adenomas has been repeatedly confirmed. The Sheffield series includes ninety patients treated from 1940 to 1960. Twenty-eight had surgery alone, forty-six surgery and radiotherapy, and sixteen radiotherapy alone. The last group consisted of three with very small tumours, five with very large tumours unsafe for surgery and eight patients who were considered unfit for operation (Emanuel, 1966) (Table 8). Analysis of the 4-year recurrence-free rate suggests that radiotherapy is at least as important as surgery in controlling these tumours. There were no complications attributable to radiotherapy: the dose given was 4000 r in 4 weeks.

Eosinophilic adenomas
There is no doubt that radiotherapy can relieve pressure effects, and it is usual for visual defects and headaches to be relieved after treatment (Hamwi, Skillman and Tufts, 1960). Unfortunately, there is no evidence

TABLE 8

Chromophobe adenoma, 1940–60 (Sheffield) (Emmanuel, 1966)

	Primary treatment	*Post-operative death*	*Recurrence free at 4 years*	%
Surgery	28	8	6	21
Radiotherapy	16	—	12	75
Surgery + post-operative radiotherapy	46	5	38	83

that any form of treatment in acromegaly prolongs life. Death is usually due to the systemic consequences of the endocrine disorder, leading to progressive disease, commonly vascular disease and diabetes, in spite of arrest of the acromegaly. The outlook is probably made worse because of the insidious nature of the disease. Most patients delay for years before attending, and as a rule they remain under medical management for a long time before they are considered for radiotherapy. This is particularly tragic when it is appreciated that the pituitary tumour is a relatively simple tumour to treat with modern radiotherapeutic techniques and that a dose of 4000 r in 4 weeks is below the tolerance limit of the surrounding normal brain tissue.

Basophilic adenomas
The nature of these so-called adenomas remains obscure. Pressure signs do not occur and a response to radiotherapy can be assessed only by a reduction of its hormonal effects. Obviously, before embarking upon a course of radiotherapy, every possible precaution should be taken to

exclude primary adrenocortical hyperplasia and adrenal tumours. In their absence, irradiation of the pituitary gland is frequently followed by regression of the stigmata of Cushing's disease and irradiation to a tumour dose of 4000 r in 4 weeks can be safely administered with reasonable optimism.

In general, radiotherapy is the treatment of choice in the management of pituitary adenomas when clinical manifestations are limited to hormonal changes or when there are only minimal signs of expansion of the sella turcica and pressure on the optic tracts. During radiotherapy regular examination of the visual fields is essential and any evidence of increase in pressure is an indication for urgent surgical decompression. When there are moderate or severe signs of pressure initially, surgery should precede radiotherapy in order to produce immediate relief of the pressure effects.

Craniopharyngiomas

Presenting in children or young adults, the characteristic feature of the tumours is slow expansion over a period of years. The tumours are usually large and are adherent to adjacent structures when first diagnosed. Although they are not extremely radiosensitive, surgical excision is usually incomplete and recurrence the rule. Radiotherapy is therefore indicated. The dose must be larger than that advised for pituitary tumours, but this appears to be fully justified by an improvement in long-term results. Kramer (1963) recommends 5500 r in 5 weeks for children and a higher dose in adults: nine out of ten of his patients were alive and well 10 years later (Kramer, 1963; Kramer, McKissock and Concannon, 1961).

Intracranial metastases

Tumours may spread into the skull, by direct invasion as is seen particularly in nasopharyngeal carcinoma, by extradural extension as is seen most frequently with malignant lymphomas and most commonly of all by the blood stream from distant primary sites.

Blood-borne cerebral metastases are extremely common. Richards and McKissock (1963) reported a series of 389 cases at one hospital in 15 years: this was 10 per cent of all the intracranial neoplasms seen. Of the 389, the primary tumour in 252 patients was a bronchial carcinoma. Our own small series of fifty-seven patients treated by radiotherapy includes thirty with bronchial carcinoma, and all other reported series confirm the high incidence of cerebral metastases from lung cancer. As this form of cancer is on the increase, cerebral metastases will be seen even more frequently in the future.

The management of patients with cerebral metastases requires a different medical approach to tumours arising in and apparently localized to the brain, when an attempt at complete cure is the usual aim. Solitary cerebral metastases are extremely rare: usually metastases are multiple and are associated with metastases elsewhere in the body. They are more likely to be truly solitary if the primary cancer was well differentiated and of low malignancy, and if there has been a long interval between apparent cure of the primary tumour and the appearance of the cerebral deposit. If there is good reason to think that a cerebral metastasis is solitary and is from a primary of low malignancy, surgical removal should be seriously considered. But surgical excision must be advised with extreme caution: an immediate mortality rate of 30 per cent or more seems an adequate deterrent (Richards and McKissock, 1963). In addition, even if a patient survives the operation, he will have to spend several weeks in hospital— which may represent a large proportion of his remaining life. If these points are considered, the long-term results of surgical removal are not impressive. Richards and McKissock reported 389 patients, of whom twenty-two survived for 1 year or more—just under 6 per cent. Four of the ten cases with a primary bronchial carcinoma also had post-operative radiotherapy.

Radiotherapy can achieve results which are as good as, or better than, those of surgery with less distress to the patient (Chao, Philips and Nickson, 1954). In our own series of fifty-seven patients treated by radiotherapy alone we have twelve alive at 1 year (21 per cent) and six have lived for 2 years or more. Radiotherapy causes no immediate mortality, and if the patients are well enough, treatment can be given as an outpatient. Recovery may commence within a few days of starting radiotherapy: obviously the longer the delay, the more permanent the damage to the brain and the smaller the chance of complete recovery.

Long-term survival after radiotherapy is possible. Bouchard and Pierce (1960) reported twenty-nine patients treated by radiotherapy—the longest survivals in this group were 51, 86 and 102 months. It would appear that radiotherapy should be considered in all patients with cerebral metastases, but obviously not used when this complication is just one manifestation of widespread clinical disease. Surgery is indicated only when there is every reason to believe that a metastasis is solitary and of low-grade malignancy.

REFERENCES

ARBIT J. (1964) In *Response of the Nervous System to Ionizing Radiation*, p. 639. Ed. HALEY T.J. and SNIDER R.S. Little, Brown & Co., U.S.A.

ARNOLD A., BAILEY P. and HARVEY R.A. (1954) *Neurology (Minneap.)* **4**, 575

ARNOLD A., BAILEY P., HARVEY R.A., HAAS L.L. and LAUGHLIN J.S. (1954) *Radiology* **62**, 37

BAILEY P. (1948) In *Intracranial Tumours*, 2nd ed. Thomas, Springfield

BEGLERE M. (1909) *Bull. Mém. Soc. méd. Hôp. (Paris)* **26**, 274

BERGER E. and ELVIDGE A.R. (1962) In *Proceedings of the 4th International Congress of Neuropathology, Munich, 1961.* Ed. JACOBS H. Georg Thieme Verlag, Stuttgart

BODEN G. (1948) *Br. J. Radiol.* **21**, 464

BODEN G. (1950) *J. Fac. Radiol.* **2**, 79

BOUCHARD J. (1966) In *Radiation Therapy of Tumours and Diseases of the Nervous System.* Lea & Febiger, Philadelphia

BOUCHARD J. and PIERCE C.B. (1960) *Am. J. Roentg.* **84**, 610

CHAO J.H., PHILIPS R. and NICKSON J.J. (1954) *Cancer* **7**, 682

CLEMENTS C.D. and HOLST E.A. (1954) *A.M.A. Archs Neurol. Psychiatry* **71**, 66

CUTLER E.C., SOSMAN M.C. and VAUGHAN W.W. (1936) *Am. J. Roentg.* **35**, 429

DOTT C.G., BAILEY P. and CUSHING H. (1925) *Br. J. Surgery* **13**, 314

DUGGER G.S., STRATFORD J.G. and BOUCHARD J. (1954) *Am. J. Roentg.* **72**, 953

DYKE C.G. and DAVIDOFF L.M. (1942) In *Roentgen Treatment of Diseases of the Nervous System.* Lea & Febiger, Philadelphia

DYNES J.B. and SMEDAL M.I. (1960) *Am. J. Roentg.* **83**, 78

ELVIDGE A., PENFIELD W. and CONE W. (1935) *Res. Publs Ass. Res. nerv. ment. Dis.* **16**, 107

EMMANUEL I.G. (1966) *Clin. Radiol.* **17**, 154

FOLTZE E.J., HOLYCKE J.B. and HEYL H.L. (1953) *J. Neurosurg.* **10**, 423

FRANKEL S.A. and GERMAN W.J. (1958) *J. Neurosurg.* **15**, 489

GRAMEGNA A. (1909) *Revue neurol.* **17**, 15

GREENFIELD M.M. and SLARK F.M. (1948) *Am. J. Roentg.* **60**, 617

HAMWI G.J., SKILLMAN T.G. and TUFTS K.C. (1960) *Am. J. Med.* **29**, 690

HENDERSON W.R. (1939) *Br. J. Surgery* **26**, 811

KAPLAN I.I. (1941) *Radiology* **36**, 588

KELLY K.H., FELSTED E.T., BROWN R.F., ORTEGA P., BIERMANN H.R., LOW-BEER B.V.A. and SHIMKEN M.G. (1951) *J. Nat. Cancer Inst.* **11**, 967

KERNOHAN J.W., MABON R.F., SVIEN H.J. and ADSON A.W. (1949) *Proceedings of a Staff Meeting, Mayo Clinic,* **24**, 71

KERR F.W.L., SCHWARTZ H.G. and SEAMAN W.B. (1954) *A.M.A. Archs Surg.* **69**, 694

KINNELDORF D.J. and HUNT E.L. (1964) In *Responses of the Nervous System to Ionizing Radiation,* p. 652. Ed. HALEY T.J. and SNIDER R.S. Little, Brown & Co., U.S.A.

KRAMER S. (1963) *Radiation Therapy of Brain Tumours: Course No. 5.* Radiological Society of North America.

KRAMER S., McKISSOCK W. and CONCANNON J.P. (1961) *J. Neurosurg.* **18**, 217

KRICHEFF I.I., BECKER M., SCHNECK S.A. and TAUERAS J.M. (1964) *Am. J. Roentg.* **91**, 167

LEWIS C.L. (1953) *Proc. R. Soc. Med.* **46**, 653

LEVY L.F. and ELVIDGE A.R. (1956) *J. Neurosurg.* **13**, 413

LINDGREN M. (1958) *Acta Radiol.* Supplement 170

MACARTY G.S. (1962) In *The Biology and Treatment of Intracranial Tumours.* Ed. FIELDS W.S. and SHARKEY P.C. C.C.Thomas, Springfield, Illinois

MALAMUD N., BOLDREY E.B., WELCH W.K. and FADELL E.J. (1954) *J. Neurosurg.* **11,** 353

MOSS W.T. (1965) In *Therapeutic Radiology, Radional, Technique, Results,* 2nd ed. C.V.Mosby, St. Louis

PATERSON R. (1963) In *The Treatment of Malignant Disease by Radiotherapy,* 2nd ed. Edward Arnold Ltd, London

PENNYBACKER J. and RUSSELL D.S. (1948) *J. Neurol. Neurosurg. Psychiat.* **11,** 183

PIERCE C.B., CONE W.V., ELVIDGE A.E. and TYE J.G. (1945) *Radiology* **45,** 247

REDMUND D.E., RINDERKNECHT R.H. and HUDGINS P.T. (1967) *Radiology* **89,** 727

RICHARDS P. and McKISSOCK W. (1963) *Brit. med. J.* **i,** 15

RICHMOND J.J. (1953) *J. Fac. Radiol.* **4,** 180

RINGERTZ N. and RAYMOND A. (1949) *J. Neuropath. exp. Neurol.* **8,** 355

ROSS J.A.T., LEAVITT S.R., HOLST E.A. and CLEMENTE C.D. (1954) *A.M.A. Archs Neurol. Psychiatry* **71,** 238

ROTH J.G. and ELVIDGE R. (1960) *J. Neurosurg.* **17,** 736

RUBENFIELD S. and ZEITEL B.E. (1956) *Am. J. Roentg.* **76,** 367

RYALL R. and JELLIFFE A.M. awaiting publication

SAMS C.F., AIRD R.B., ADAMS G.D. and ELLIMANS G.L. (1964) In *Response of the Nervous System to Ionizing Radiation,* p. 554. Ed. HALEY T.J. and SNIDER R.S. Little, Brown & Co., U.S.A.

SCHOLTZ N. and HSU Y.K. (1938) *Archs Neurol. Psychiatry* **40,** 928

SHELINE G.E., BOLDREY E., KARLSBERG P. and PHILLIPS T.L. (1964) *Radiology* **82,** 84

SMITH R.A., LAMPE I. and KAHN E.H. (1961) *J. Neurosurg.* **18,** 91

SMITHERS D.W., CLARKSON J.R. and STRONG J.A. (1943) *Am. J. Roentg.* **49,** 606

STEVENSON L.D. and ECKHARDT R.E. (1945) *Archs Path.* **39,** 109

SVIEN H.J., GATES E.M. and KERNOHAN J.W. (1949) *Archs Neurol. Psychiatry* **62,** 847

TAVERAS J.M., THOMPSON H.G. and POOL J.L. (1962) *Am. J. Roentg.* **87,** 473

TOBIAS C.A., LAWRENCE J.H., LYMAN J., BORN J.L., GOTTSCHALK A., LINFOOT J. and MACDONALD J. (1964) In *Responses of the Nervous System to Ionizing Irradiation,* p. 19. Ed. HALEY T.J. and SNIDER R.S. Little, Brown & Co., U.S.A.

VAETH J. (1965) In *Progress in Radiation Therapy,* Vol. III, p. 16. Ed. BUSCHKE F. Grune & Stratton, N.Y.

WACHOWSKY T.J. and CHENAULT H. (1945) *Radiology* **45,** 227

ZELENY V. (1956) In *Progress in Radiobiology,* p. 403. Ed. MITCHELL J.S., HOLMES B.E. and SMITH C.L. Oliver & Boyd, Edinburgh

ZEMAN W. (1949) *Arch. Psychiat. NervKrankh.* **182,** 713

ZIMMERMAN A.M., NETSKY M.G. and DAVIDOFF L.M. (1956) In *Atlas of Tumours of the Nervous System.* Lea & Febiger, Philadelphia

CHAPTER FOURTEEN

BIOCHEMICAL ASPECTS OF
IONIZING RADIATION IN NEUROLOGY

P.T.LASCELLES

Since the discovery of X-rays in 1895 ionizing radiation has been increasingly employed in neurology for diagnostic, therapeutic and research purposes. Fig. 1 shows the evolution of the diagnostic techniques.

Radioactive isotopes have been used very extensively for research purposes in the basic sciences related to neurology for many years, and any attempt to review comprehensively the part they have played would be a formidable task. Their application in the diagnostic field, too, has grown enormously since the advent of brain scanning following the pioneer work which was carried out in the late 1940s and early 1950s. The almost explosive increase in the application of radioactive isotopes to brain and spinal cord scanning in the last few years can be attributed to the vastly superior and sophisticated detecting and counting equipment which is now produced commercially at an acceptable price, together with the availability of more suitable isotopes. Berrocal (1963a, b) has fully reviewed the isotopes available for brain scanning, with a critical assessment of their relative merits. Brain scanning has now been adopted as a routine technique which is both safe and makes little demand from patients.

Radiotherapy, too, is being increasingly employed in the treatment of neurological disease, particularly since the introduction of megavoltage techniques. This means that ionizing radiations of widely differing energy values are being used for the investigation and treatment of neurological disorders.

THE RESPONSE OF THE CELL AND ITS
COMPONENTS TO IONIZING RADIATION

This is best reviewed in terms of general principles applying to any tissue. The special features of nerve tissue can then be considered against this background.

When X-rays enter cells they largely pass through unaltered, especially the shorter wavelengths. However, a certain amount of energy is dispersed, resulting in the liberation of high-energy electrons. These electrons

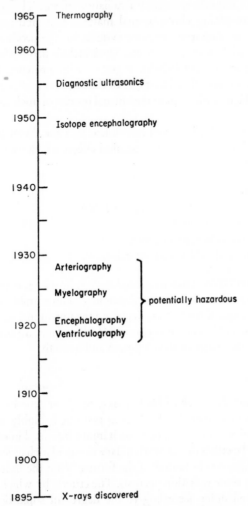

FIG. 1. Scheme showing evolution of radiological diagnostic procedures. (Reproduced by permission from Bull, 1966.)

produce excitation and ionization along their tracks and at points irregularly scattered close to the atoms from which each electron was ejected. The degree of radiation damage produced depends upon the type of radiation, the absorbed dose and the physiological state of the cells at the time of exposure. Greater damage is caused in an actively metabolizing cell, particularly during mitosis. This in itself suggests that the reproductive apparatus is an important target for radiation effects. The cell nucleus in addition to containing chromosomal material is a vital synthetic locus containing many important enzyme systems for the production of both nuclear and cytoplasmic components. These include deoxyribonucleic acid (DNA), ribonucleic acid (RNA), together with proteins, histones and lipids. It has been postulated that one of the important effects of radiation on the cell is the modification of the normal process of nuclear-cytoplasmic exchange.

On the other hand, it is well established that the direct irradiation of DNA *in vitro* has certain specific chemical effects which may be summarized as follows:

1   Change or loss of base from DNA.
2   Single and double chain fracture of DNA.
3   Polymerization of DNA.
4   Thiamine cross-linkage in DNA.
5   Combination of DNA with proteins.

The effect on DNA when irradiated *in vivo*, however, is probably very different and much more complex depending upon a multitude of factors, including the stage of the mitotic cycle and the nutritional state of the cell. As far as the nervous system is concerned, these facts are mainly of relevance during the stages of development and growth.

The neurone

Largely through the work of Hicks (1953, 1954) on experimental animals it has become clear that the developing neurone is highly susceptible to ionizing radiation, but resistance to such injury increased in older animals. Damage may be either direct or secondary to vascular changes, though the weight of evidence is in favour of the former being the more important, particularly in acute radiation necrosis. The criteria by which cell damage may be assessed differ according to the degree of sophistication of the methods of observation. A cell which is not able to replicate by virtue of radiation damage is frequently capable of carrying on other biochemical functions apparently normally for considerable periods. Light microscopic changes are not striking in the acute phase of radiation reaction to X-rays, though Alvord and Brace (1957) and others have described pyknosis of granular cells in the cerebellum which they considered to coincide with a period of neurological dysfunction. Campbell and Novick (1949) described the direct effect on nerve cells of β radiation.

Sensitivity to radiation varies greatly in different areas of the brain and most workers agree that the brain stem is the most sensitive region (Bailey, 1962). In the later stages of the reaction to radiation there is extensive damage to neurones with a selectivity for the white matter, together with degenerative changes in blood vessels.

Electron microscopic studies (Pitcock, 1962) have confirmed the clumping of nuclear chromatin in cerebellar granular cells. This clumping was found to vary markedly from cell to cell. Radiation also produced characteristic changes in the nuclear membrane of these cells.

Some experimental work has also been carried out on the response of granule cells to very high dosages of γ radiation (Vogel, 1962). The main effect was to induce a marked hypertonicity of the cytoplasm which caused transient and reversible light microscopic changes in rabbit brain but rapid cellular death in dogs.

The glia

Oligodendroglial cells are affected markedly and early after X-irradiation and undergo swelling of cytoplasm (Arnold and Bailey, 1954), followed by pyknosis and disintegration of cells (Hicks, Leigh and Wright, 1956). Arnold and Bailey (1954) reported a reduced formation of compound granular corpuscles in areas of radiation necrosis.

The response of astrocytes is also marked, particularly to β radiation (Campbell and Novick, 1949). Initially there is swelling and fragmentation of cells followed by a disappearance of astrocytes from the area of radiation necrosis. Only after several months is there a resumption of the normal astrocytic proliferation in repair of necrotic areas.

Electron microscopy revealed changes in cells which were not shown by light microscopy (Pitcock, 1962). Nuclear 'blebbing' is observed frequently, mitochondrial swelling is sometimes evident and changes in endoplasmic reticulum are described together with occasional swelling of endothelial cells of capillaries.

Synapses

Clemente and Holst (1954) reported that the permeability of the neuronal membrane increases after acute irradiation. Generally the ionic fluxes across a cell membrane are driven by two forces: (a) the passive diffusion gradient due to the difference in concentration across the membrane and

(b) the active 'pump' mechanism against the diffusion gradient for which energy is supplied by cellular metabolism.

Bresciani, Auricchio and Fiore (1962), have shown that the increase in sodium content of irradiated human erythrocytes following exposure to ionizing radiation is mainly due to depressed active sodium efflux. The passive sodium fluxes also increase gradually to augment the intracellular sodium content. They emphasized the higher radiosensitivity of the sodium pump mechanism in comparison with passive fluxes. If the same principles applied to the neuronal membrane, the membrane resting potential (MRP) should decrease until intracellular potassium decreased to compensate for the excess cation. The amplitude of intracellular spike response (ISR) should also be impaired after radiation on account of decreased diffusional gradient for sodium across the membrane.

Experimentally, however, the MRP showed no change even after high doses of irradiation, and the ISR increased. Thus there is clearly a different biochemical response between the two types of cells. It is probable that in the case of the neuronal membrane both passive efflux and influx of sodium may increase after irradiation to the same degree, so that the MRP remains constant with the aid of the active sodium pump which is radioresistant.

Membrane permeability increase during excitatory post-synaptic potential (EPSP) is ionically non-specific, while that during inhibitory post-synaptic potention (IPSP) is specific for certain ions, for example, potassium, chloride, bromide (Eccles, 1957). Thus if irradiation specifically affects sodium permeability its effects would be manifested mainly on EPSP rather than IPSP and this is in accordance with experimental results.

Blood-brain barrier

The concept of the blood-brain barrier originates from Ehrlich's observations that acid aniline dyes when injected into animals fail to stain the living brain while at the same time permeating other tissues. Many other substances, including electrolytes, colloids and more complex molecules behave similarly, though it may be mentioned that basic dyes do generally penetrate the central nervous system readily.

Rachmanow (1926) first demonstrated that irradiation was followed by the appearance of blue granules in the mesodermal tissues of the brain of animals previously injected with trypan blue. No staining of ecto-dermal elements was observed, however.

For many years investigators made unsuccessful attempts to identify the barrier anatomically. When radioactive isotopes became available it was found that many of them penetrated into the brain at a very slow rate. Much of this experimental work is reviewed in the monograph by Bakay (1956).

In 1954, Clemente and Holst found local alterations of the barrier for trypan blue after irradiation of the heads of monkeys, the dorsal medulla and hypothalamus showing the most marked blue discoloration. Microscopic examination revealed degeneration of the membranes around the blood vessels and neuronal degeneration. These changes correlated well with the regions of the brain that showed trypan blue staining.

Electron microscopy has demonstrated that extracellular space in the brain is exceedingly small. Within the central nervous system no extracellular space surrounds the capillaries. Glial processes both astrocytic and oligodendroglial abut directly on the basement membrane of the capillary endothelium. The consequence of this unique lack of interstitial space is that transport mechanisms must be mainly through glial cells rather than through an extracellular space.

Electron microscopic studies of irradiated nervous tissue are not numerous. Clemente and Richardson in 1962 reviewed the studies dealing with the influence of X-rays on the blood-brain barrier and the associated changes in blood vessels of the central nervous system after irradiation. Larsson (1960), studied the lesions produced in the rat brain by high-energy protons with particular reference to vascular changes. The first effects observed were impaired capillary circulation and damage to the blood-brain barrier for trypan blue. These changes were succeeded by degeneration of nerve cells leading to necrosis. The vascular disturbances seemed to precede the trypan blue staining, and were probably the direct cause of damage to the parenchyma.

Thus the changes in the blood-brain barrier caused by ionizing radiation represent fundamental alterations with corresponding biochemical connotations.

The concept of the blood-brain barrier has been the subject of renewed interest from an entirely different point of view. Since the introduction of radioactive isotopes for brain and spinal cord scanning it has been found that a wide variety of pathological lesions affecting the central nervous system take up the isotope more readily than the surrounding healthy tissue. The process of uptake appears to be non-specific and is regarded as a passive diffusion of the isotope into the lesion as a result of a local breakdown of the blood-brain barrier caused by the pathological process.

Only rarely is a 'negative' uptake encountered as, for example, in an infarcted area following thrombosis of a major vessel.

Some attempts have been made to elucidate in greater detail the mode of uptake of the isotope and its subcellular distribution.

Raimondi (1964) gave radio-iodinated serum albumin to two patients with astrocytomas 20 hours prior to surgery. Specimens of the tumour were examined electron microscopically after removal.

It was concluded from this study that the labelled protein was transferred into the capillary endothelium by pinocytosis and that the degree of pinocytotic activity was markedly increased in the neoplastic cells.

Nevertheless, the constancy with which particularly undifferentiated tumours take up a large number of non-protein-bound isotopes suggest that the mechanism of uptake must be non-specific, a certain proportion of radioactivity being found in oedema fluid and haemorrhage surrounding pathological lesions. Moreover, the pattern of uptake with time is specific for some pathological processes, for example cerebral abscess, and repeated scans over a period of hours and the Planiol technique is currently being used as a method of assessing the nature of the pathological process.

Cerebral oedema

Howell, Embree and Tatlow (1963) described a reactive oedema in the host brain tissue surrounding neoplasms. It has been suggested that such oedema may be further increased by irradiation, and echo encephalographic studies have indicated that there is a shift of midline structures attributed to cerebral oedema following radiotherapy to one hemisphere, particularly during the early phases of treatment. In patients who received large doses of steroids during the early phase of radiotherapy there was a return of the midline structures to the normal position. When steroid treatment was discontinued, evidence of midline shift returned (Son, unpublished).

MISCELLANEOUS EFFECTS OF RADIATION ON BRAIN BIOCHEMISTRY

Histochemical studies

The histochemical changes secondary to irradiation follow the general pattern of tissue responses by following into the two categories of immediate direct injury of cells and delayed indirect injury.

Brownson, Suter and Diller (1963) stressed the radiosensitivity of the oligodendrocyte and, with the exception of the granule cells of the cerebellum, the relative radioresistance of the neurones. In the oligodendrocyte the most conspicuous acute changes detected by light microscopy were in the nuclear chromatin. It was found that DNA underwent a change that rendered the molecule resistant to DNAase hydrolysis.

Zeman and Curtis (1962) found that acute irradiation had no effect on oxidative enzymes in the central nervous system unless actual cell necrosis was produced.

Zeman (1961, 1962a, b) also reported resistance of oxidative enzymes to chronic radiation effects.

Brownson, Suter, Oliver, Ingersoll and Burt (1963) made a study of histochemical changes in rats in an attempt to correlate both acute and chronic phases of radiation-induced changes in the central nervous system. Histochemical procedures for acid phosphatase indicated that the activity of this enzyme responded to X-irradiation of 5000 r or higher, as manifested by enlargement or clumping of acid phosphatase-positive granules within the cells, maximal at about 48 hours following irradiation, and declining over a period of the next week. 2500 r or less produced no demonstrable effect at any time.

Succinic dehydrogenase activity in neurones was altered as compared to controls, the changes being evident at 6 hours, rising to a peak at 12 hours and subsiding at 48 hours.

The authors conclude that the increased enzyme activity was indicative of a response to radiation in the region of the capillary wall and that this was probably related to the alteration in the blood-brain barrier resulting from the ionizing radiation.

Autoradiographic studies
This is an investigational technique which can provide data of metabolic processes on a cellular and subcellular level, which is difficult to obtain by other methods. Preliminary work indicated that it may be particularly useful for the investigation of DNA, RNA and protein metabolism, but lipid chemistry, electrolyte and hormonal changes can also be studied by this means. The chief disadvantage of this technique is that it is a time-consuming procedure. Altman (1964) has carried out studies of DNA metabolism in normal and pathological cell multiplication.

It is known (Taylor, Woods and Hughes, 1957; Hughes, Bond, Brecher, Cronkite, Painter, Quastler and Sherman, 1958) that cells which prepare for mitosis in a medium in which thymidine-^{3}H is present, and only such cells, will utilize this specific precursor of DNA for duplication of chromosomes.

Another study by Altman (1962a) suggested that not only microglia cells but also astrocytes and particularly oligodendrocytic cells multiplied following brain lesions. He suggested that since these cell types are believed to be involved in the myelinization of nerves their multiplication indicates the possibility of regeneration processes in the brain following pathological lesions.

He also found (Altman, 1962b) labelling of nuclei of occasional neurones in forebrain structures not necessarily associated with pathological lesions. This was a most surprising observation in view of the fact that neurones are not generally considered to divide after embryonic development. As the authors state, however, further studies are required to confirm these findings and to exclude the possibility of artefacts.

TABLE I

Oxygen uptake and carbon dioxide production in cortex and hypothalamus of intact control, irradiated and castrated rats

Procedure	Oxygen uptake (μl O_2/mg wet tissue/hr)		Carbon dioxide production (μl CO_2/mg wet tissue/hr)	
	Cortex	Hypothalamus	Cortex	Hypothalamus
Controls	0.60 ± 0.03*	1.09 ± 0.09	0.97 ± 0.12	1.21 ± 0.16
Irradiated†	0.95 ± 0.08	0.86 ± 0.09	0.30 ± 0.06	1.12 ± 0.10
	(<0.01)‡		(<0.01)	
Castrated§	0.53 ± 0.05	0.63 ± 0.06	0.89 ± 0.05	0.67 ± 0.19
		(<0.01)		(<0.05)

* Mean ± standard error.

† 500 r whole-body X-irradiation at 2 days. Animals sacrificed at approximately 75 days.

‡ Figures in parentheses are p values for differences between experimental and control groups.

§ Rats castrated at 5 days and sacrificed at approximately 75 days.

RESPIRATION STUDIES OF RAT BRAIN WITH AND WITHOUT X-IRRADIATION

Timiras, Woodbury and Goodman (1954), using the Warburg technique, made measurements of oxygen consumption and carbon dioxide production by rat tissues with and without whole body X-irradiation (500 r

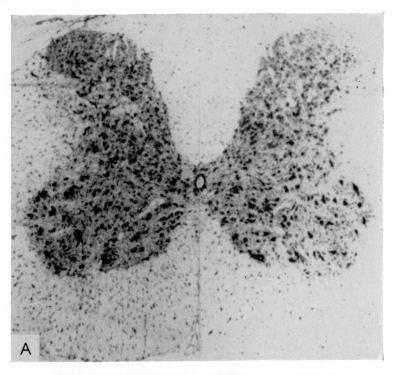

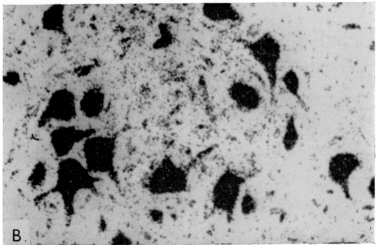

FIG. 2.(a) Autoradiogram of a section of the spinal cord of a rat injected intraperitoneally with 2 mc of ³H-leucine and killed 2 hours after injection. 5 months' exposure × 30. (b) Ventral horn cells at higher magnification.

in a single dose on the second day after birth). Tissue respiration studies were made 3 months later.

It was found that radiation significantly increased the oxygen consumption of cerebral cortex and aorta, but not of hypothalamic tissue. Carbon dioxide production was decreased in the cerebral cortex and to a less extent in the aorta. Hypothalamic carbon dioxide production was unaffected.

Adrenal hypertrophy and testicular atrophy were noted in the irradiated rats, and the author postulated that the effects of radiation on cerebral aerobic and anaerobic metabolism might be directly related to the endocrine changes. However, experiments with rats castrated or irradiated at an early age indicated that radiation affected mainly the cortex and castration the hypothalamus. Some of the details of these results are seen in Table 1.

EFFECTS OF IONIZING RADIATION ON LIPID CHEMISTRY

No review of the effects of ionizing radiation on brain biochemistry would be complete without a reference to the changes in lipid chemistry. Although a number of papers have been devoted to this subject, the changes described on the whole have been minor.

Coniglio, Davis, Windler and Tsung (1964) studied the biochemical changes in brain lipids 24 hours after a 10,000 r dose of total-body γ irradiation to newborn and weanling rats. They found a slightly decreased total fatty acid concentration in the former and increased palmitic and palmitoleic acid concentration in the latter. Other changes were noted but were not regarded as definitely significant. No differences were noted in phospholipid content between controls and tests.

Grossi, Poggi and Paoletti (1964) reviewed the effect of ionizing radiations on lipid synthesis in the brain and liver of animals, and the influence upon this of radiosensitizing and radioprotective drugs. Cysteamine protected the liver and brain of irradiated rats against the known increased and decreased rate in each organ respectively of lipid biosynthesis from acetate. An opposite effect was found with Synkayvite when administered to animals submitted to 2000 r X-irradiation which induced a further increase of acetate incorporation into liver cholesterol and fatty acid and a less important further decrease of the precursor incorporation into brain lipids.

The mechanism of these changes was further investigated and it was concluded that the increased liver lipid synthesis was a reaction to an impairment of the normal exchanges between liver and plasma cholesterol, but that no satisfactory explanation has been found for the decreased levels of brain lipid synthesis.

EFFECT OF IONIZING RADIATION ON BRAIN LYSOSOMES

Bacq and Alexander (1961a) have shown that in irradiated tissues there is an increased activity of certain enzymes including acid hydrolases, and, as mentioned above, acid phosphatase activity in irradiated neuronal tissues is also enhanced. The mechanism of this activation of lysosomal activity is unknown, but Bacq and Alexander (1961b) have proposed that it is due to changes in permeability of intracellular membranes. This, however, does not explain the failures of X-rays to activate enzymes *in vitro* (Bacq and Alexander, 1961b). Koenig (1964) suggests that nuclear histones liberated by X-irradiation from protein (Meisel, Brumberg, Kondritjev and Barsky, 1961) migrate into the cytoplasm and bind with lysosomes, causing an excessive release of hydrolytic enzymes.

IONIZING RADIATION AND THE RETINA

Lebedinskiy and Nakhil'Nitskaya (1963) have reviewed a series of papers on the biochemical and electrophysiological responses of the retina to ionizing radiation.

It has been known for many years that visual purple is resistant to high doses of X-rays, no apparent bleaching being produced *in vitro* with doses as high as 300 kr. According to Peskin (1955), this resistance is due to the −SH group being inaccessible to the action of small quantities of peroxides formed in the tissue during irradiation.

Lipetz (1955a, b) studied the effect of high-energy heavy-particle irradiation of the retina, but again bleaching of visual purple was only demonstrated at doses which produced tissue destruction probably caused by local heating effects. Experimentally, X-rays also produce bleaching of the retina not attributable to heating, but the dose required for this is enormous (10,000 kr).

Noell (1962) studied the effect of X-irradiation on visual cells, particularly rod cells. He concluded that there was a close association of visual cell death, sensory function and energy-yielding metabolic activities. It was found that all aspects of cell function failed simultaneously with the same dose of irradiation.

The hypothesis that a particular part of the cell, for example, the nucleus or the mitochondria, might be the critical target for X-irradiation, disorganization of which would disrupt all functions of the cell, does not bear critical examination. Detailed studies of electrical function have been made (Gaffey, 1964), and it has been shown that these and respiratory activity are not closely related to the integrity of any particular part of the cell. The mechanism of disruption of cell activity, therefore, by irradiation must take account of the disparate aspect of function in these cells.

This review of the biochemical aspects of ionizing radiation in relation to neurology has necessarily been brief and highly selective. This is an enormous field of scientific inquiry which essentially requires a multi-disciplinary approach for its exploration.

No mention has been made of the functional changes brought about by the irradiation of brain, either at the physiological or behavioural level, nor has mention been made of the pharmacological aspects of irradiated nerve.

An aspect of the problem which is receiving great attention at the present time is the effect of ionization due to high-energy elementary particle bombardment of the nervous system on account of its importance in relation to space travel.

These problems, however, still wait detailed biochemical elucidation.

The figure and tables are reproduced by the kindness of Dr E.G. Paoletti, whom I thank for this permission.

REFERENCES

ALTMAN J. (1962a) *Exp. Neurol.* **5,** 302

ALTMAN J. (1962b) *Science* **135,** 1127

ALVORD E.C. and BRACE K.C. (1957) *J. Neuropath. Exp. Neurol.* **16,** 3

ARNOLD A. and BAILEY P. (1954) *A.M.A. Arch. Pathol.* **57,** 383

BACQ Z.M. and ALEXANDER P. (1961a) In *Fundamentals of Radiobiology,* 2nd ed., p. 332. Pergamon Press, New York and London

BACQ Z.M. and ALEXANDER P. (1961b) In *Fundamentals of Radiobiology,* 2nd ed., p. 274. Pergamon Press, New York and London

BAILEY O.T. (1962) In *Response of the Nervous System to Ionizing Radiation,* p. 165. Ed. HALEY T.J. and SNIDER R.S., Academic Press, New York and London

BAKAY L. (1959) In *The Blood-brain Barrier with Special Regard to the Use of Radioactive Isotopes.* Charles C.Thomas, Springfield, Illinois.

BERROCAL J.O. (1963a) *Revta. clin. esp.* **91,** 1

BERROCAL J.O. (1963b) *Revta. clin. esp.* **91,** 333

BRESCIANI F., AURICCHIO F. and FIORE C. (1962) *Biochem. Biophys. Res. Commun.* **8,** 374

P

BROWNSON R.H., SUTER D.B. and DILLER D.A. (1963) *Proc. Soc. Exp. Biol. Med.*, *N.Y.* **55**, 243

BROWNSON R.H., SUTER D.B., OLIVER J.L., INGERSOLL E.H. and BURT D.H. (1963) In *Response of the Nervous System to Ionizing Radiation*, p. 307. Ed. HALEY T.J. and SNIDER R.S. J. & A. Churchill Ltd, London

BULL J.W.D. (1966) *J. Roy. Coll. Phycns. Lond.* **1**, 75

CAMPBELL B. and NOVICK R. (1949) *Proc. Soc. Exp. Biol. Med.*, *N.Y.* **72**, 34

CLEMENTE C.D. and HOLST E.A. (1954) *A.M.A. Arch. Neurol. Psychiatry* **71**, 66

CLEMENTE C.D. and RICHARDSON H.E. Jr (1962) In *Response of the Nervous System to Ionizing Radiation*, p. 411. Ed. HALEY T.J. and SNIDER R.S. Academic Press Inc., New York

CONIGLIO J.G., DAVIS J.T., WINDLER F. and TSUNG V. (1964) In *Response of the Nervous System to Ionizing Radiation*, p. 377. Ed. HALEY T.J. and SNIDER R.S. J. & A. Churchill Ltd, London

ECCLES J.C. (1957) In *The Physiology of Nerve Cells*. Johns Hopkins Press, Baltimore

GAFFEY C.T. (1964) In *Response of the Nervous System to Ionizing Radiation*, p. 243. Ed. HALEY T.J. and SNIDER R.S. J. & A. Churchill Ltd, London

GROSSI E., POGGI M. and PAOLETTI R. (1964) In *Response of the Nervous System to Ionizing Radiation*, p. 388. Ed. HALEY T.J. and SNIDER R.S. J. & A. Churchill Ltd, London

HICKS S.P. (1953) *Am. J. Roentg.* **69**, 272

HICKS S.P. (1954) *Arch. Pathol.* **57**, 363

HICKS S.P., WRIGHT K.A. and LEIGH K.E. (1956) *Arch. Pathol.* **61**, 226

HOWELL D.A., EMBREE G. and TATLOW T.W. (1963) *Can. med. Ass. J.* **89**, 866

HUGHES W.L., BOND V.P., BRECHER F., CRONKITE E.P., PAINTER R.S., QUASTLER H. and SHERMAN F.G. (1958) *Proc. Nat. Acad. Sci. U.S.A.* **44**, 476

KOENIG H. (1964) In *Response of the Nervous System to Ionizing Radiation*, p. 403. Ed. HALEY T.J. and SNIDER R.S. J. & A. Churchill Ltd, London

LARSSON B. (1960) *Acta Soc. Med. Upsalien* **65**, 404

LEBEDINSKIY A.V. and NAKHIL'NITSKAYA Z.N. (1963) In *Effects of Ionizing Radiation on the Nervous System*. Elsevier Publishing Company, London

LIPETZ L.E. (1955a) *Brit. J. Ophthamol.* **39**, 577

LIPETZ L.E. (1955b) *Radiation Research* **2**, 306

MEISEL M.N., BRUMBERG E.M., KONDRITJEVA T.A. and BARSKY I.J. (1961) In *The Initial Effects of Ionizing Radiation on Cells*, p. 107. Ed. HARRIS R.J. Academic Press, New York and London

NOELL W.K. (1962) In *Response of the Nervous System to Ionizing Radiation*, p. 543. Ed. HALEY T.J. and SNIDER R.S. Academic Press Inc., New York

PESKIN J.C. (1955) *Amer. J. Ophthalmol.* **39**, 849

PITCOCK J.A. (1962) *Lab. Invest.* **11**, 32

RACHMANOW A. (1926) *Strahlentherapie* **23**, 318

RAIMONDI A.J. (1964) *Archs Neurol., Chicago* **11**, 173

SON, Y.H. Unpublished. Personal communication

TAYLOR J.H., WOODS P.S. and HUGHES W.L. (1957) *Proc. Nat. Acad. Sci. U.S.A.* **43**, 122

TIMIRAS P.S., WOODBURY D.M. and GOODMAN L.S. (1954) *J. Pharmacol. exp. Ther.* **112**, 80

TIMIRAS P.S., MOGUILEVSKY J.A. and GEE S. (1964) In *Response of the Nervous System to Ionizing Radiation*, p. 365. Ed. HALEY T.J. and SNIDER R.S. J. & A. Churchill Ltd, London

VOGEL F.S. (1962) In *Response of the Nervous System to Ionizing Radiation*, p. 249. Ed. HALEY T.J. and SNIDER R.S. Academic Press, New York and London

ZEMAN W. (1961) In *Fundamental Aspects of Radiosensitivity*, p. 176. Ed. SPARROW A.H. Off. Tech. Serv. Dept. Commerce, Washington, D.C.

ZEMAN W. and CURTIS H.J. (1962) In *Proceedings of the 4th International Congress of Neuropathology*, p. 141. Ed. JACOB H. George Thieme Verlag, Stuttgart

ZEMAN W., CURTIS H.J., SCARPELLI D.G. and KLEINFIELD R. (1962a) In *Response of the Nervous System to Ionizing Radiation* (1st Internat. Symposium), p. 429. Ed. HALEY T.J. and SNIDER R.S. Academic Press, New York and London

ZEMAN W., SAMORAJSKI T. and CURTIS H.J. (1962b) In *Effects of Ionizing Radiation on the Nervous System*, p. 297. International Atomic Energy Agency, Vienna

CHAPTER FIFTEEN

HYDROCEPHALUS

VALENTINE LOGUE

CLASSIFICATION

The classification of the different varieties of hydrocephalus has tended to become more simplified with the passage of years. In the classification which follows I am of course excluding replacement hydrocephalus or hydrocephalus 'ex vacuo' where fluid is merely taking the place of atrophied or undeveloped brain, as well as the condition of 'external hydrocephalus' which in most cases has represented a developmental arachnoidal cyst.

With one possible exception all hydrocephalus is obstructive and the different morbid anatomical and radiological changes which may be observed derive entirely from the site of the obstruction in the CSF pathways. This is not a statement of anything new. D.S.Russell was saying this 20 years ago and many authors have reiterated it since, but it does perhaps need emphasizing.

The one possible exception to this rule is that of papilloma of the choroid plexus in which condition hydrocephalus is said to be due to over-production of CSF above a level capable of being dealt with through the normal absorptive channels, whatever their capacity may be.

PAPILLOMA OF THE CHOROID PLEXUS

I shall examine in a little more detail this question of oversecretion of CSF by papillomas with apologies for referring to a subject which has become rather hackneyed. In passing I should comment briefly on the clinical features of these papillomas. Although they represent only 0.5 per cent of all intracranial tumours, nevertheless they comprise 3 per cent of tumours in infancy and childhood. The condition presents usually after the first year of life and often with papilloedema, which is uncommon in the other forms of childhood hydrocephalus; the ventricles may not be particularly enlarged; the CSF usually has a raised protein content and may be xanthochromic, and finally, those papillomas in the lateral ventricles are commonly not diagnosed during life because they are so easily missed on the 'bubble' ventriculograms, the common method of cerebral investigation in infants. Their discovery depends on a very large air replacement with horizontal X-ray views of each of the lateral ventricles.

It would be reasonable to assume that a papilloma, which represents merely a very large aggregation of choroidal fronds which under the microscope differ only marginally from the normal choroid plexus, should be capable of producing a larger amount of CSF than the normal plexus. Some support has been lent to this clinically by a number of observations, such as that of Ray and Peck (1956), in a child of $2\frac{1}{2}$ months with a papilloma in each of the lateral ventricles in whom they carried out an arachno-ureteric shunt, which was followed by the passage of such large volumes of urine that frequent episodes of severe dehydration and electrolyte loss resulted, requiring a daily infusion of $\frac{1}{2}$ to 1 litre of half-normal saline. But excessive production does not necessarily imply inability to discharge the CSF into the venous system.

This problem should be susceptible to experimental clinical proof; for instance, in a case in which a diagnosis has been made of papilloma situated in the lateral ventricle (not in the third or fourth ventricles because in these sites mechanical obstruction of the ventricular system occurs) the following criteria should be established:

1   An air study should show the passage of air over the surface of the brain, indicating that there is no obstruction to the CSF pathways at the tentorial hiatus or over the convexity.
2   The CSF should not have an excessively raised protein content because this feature alone may produce interference with absorption.
3   The tumour should be removed through an incision directly over the ventricle across the convexity of the brain without opening up the basal cisterns.
4   Following the surgical removal the hydrocephalus should subside without further action being necessary.

Well over 400 cases have now been reported in the literature, although many of them were found unexpectedly at autopsy, but I cannot find one case in which all these criteria are satisfied. Many of the cases have not been investigated with air studies. Most of those which have been so investigated demonstrate a block at the tentorial hiatus, and in the other cases the CSF contained raised protein and in some was yellow or showed evidence of the spontaneous haemorrhage to which these tumours are liable.

It seems to me the more likely view is that the haemorrhage and high protein content of the CSF with these tumours leads after a time to adhesive block at the tentorial hiatus or above, producing an obstructive hydrocephalus.

With excessive quantities of CSF travelling through these pathways the obstruction would not need to be very great and when the papilloma was removed the pathways which remained patent would probably be quite capable of dealing with the reduced fluid production of the residual normal plexuses.

I have used some space in dealing with papillomas to show that in my view the type of hydrocephalus they produce comes within the classification that all hydrocephalus, without exception, is obstructive in nature.

CSF ABSORPTION

Professor Davson will deal in Chapter 16 with the detailed aspects of the physiology of flow and absorption. I will content myself solely with one facet, that of the function of the arachnoidal villi in the sagittal sinus. This was worked out very ingeniously by Welch and Friedman in 1960 in monkeys. They demonstrated for the first time that a villus when distended with fluid had the appearance of a convoluted mass of tubules, measuring between 4 and 12 microns in diameter with larger tubules at the base, and the openings of the tubes on to the surface being seen as pits. Their size would therefore permit the passage of intact red blood corpuscles.

The tubules have only a flattened single cell layer between them so that the whole structure is very delicate and can be blown out and distended when CSF is flowing through it, but conversely only the slightest rise of pressure of venous blood in the sagittal sinus would be needed to collapse the tubules down and obliterate their lumina. It has been shown experimentally that there is normally a pressure of 20–40 mm of water higher in the CSF than in the sagittal sinus tending to drive the fluid through the villi into the sinus. With a sudden increase of sinus pressure above that of the CSF these delicate villi would collapse down, flattening the tubules and preventing any entrance of blood from the sinus into the CSF. They act, therefore, as very simple but very effective one-way valves. Welch & Friedman confirmed the fluid would pass only in one direction, from outside to inside the sinus through them, and that the valves had a critical opening pressure of between 20 and 50 mm of CSF.

SITE OF OBSTRUCTION

The site of obstruction of CSF pathways has been of particular importance in the past when surgical treatment in the form of diversion or shunting the fluid was required. Dandy and Blackfan (1914) clarified the principles in their classic work, using dogs as the experimental animal and the dye phenolsulphonephthalein, which is absorbed diffusely from the spinal and cerebral subarachnoid spaces but not from the ventricles. They defined the terms obstructive and communicating hydrocephalus which have been in common use ever since. The features are well known. In obstructive hydrocephalus an intraventricular injection of PSP cannot be recovered from the lumbar theca and its appearance in the urine is delayed for up to 30–60 minutes and then only small quantities, 1–2 per cent of the total injected, is excreted. Whereas in communicating hydrocephalus, where the ventricles communicate with the spinal theca, the dye appears in the lumbar region within minutes and much greater quantities appear earlier in the urine, 10–15 per cent being excreted per hour in the first 2–3 hours. Dandy emphasized the surgical importance of communicating hydrocephalus because its presence indicated that as well as the ventricles, the lumbar theca could be used as the reservoir from which to divert CSF, but diversion would have to be to absorptive channels outside the central nervous system.

In obstructive hydrocephalus the short circuit could remain within the head, the CSF being diverted from the ventricles into the normal absorptive channels and he used the route from the third ventricle into the suprasellar and chiasmatic cisterns by tearing open the floor of the third ventricle—third ventriculostomy.

This distinction between obstructive and communicating hydrocephalus (I would suggest perhaps a better terminology would be, as all hydrocephalus is obstructive, ventricular and extra-ventricular hydrocephalus) has become much less important in recent years because the modern valvular shunting procedures use the ventricles routinely as the draining reservoir irrespective of the obstruction being ventricular or extra-ventricular. But the distinction becomes important again if the valvular shunt fails and one has to fall back on one of the older surgical methods that will be mentioned shortly.

OBSTRUCTION IN INFANTS AND ADULTS

The site of obstruction of the CSF pathways and their various causes differ between adults and infants.

Fig. 1, which has been modified from a drawing by Luessenhop (1966),

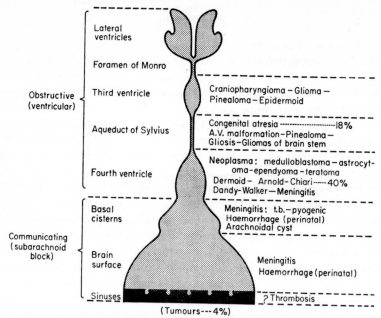

FIG. 1. Sites of obstruction in CSF pathways in children.

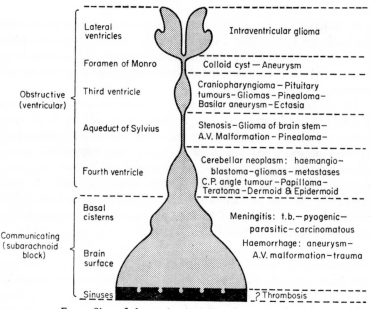

FIG. 2. Sites of obstruction in CSF pathways in adults.

illustrates in diagrammatic form the pathways of the CSF laid out in one plane reaching from the lateral ventricles anteriorly to the sagittal sinus posteriorly. It will be noted in young people that the sites of obstruction are mainly at the two isthmi, the aqueduct and the outlet of the fourth ventricle, and congential malformations form a large proportion of causes.

Fig. 2 illustrates the site of obstruction in adults and demonstrates that tumours have now become the most frequent cause of hydrocephalus and the sites of obstruction are spaced more evenly throughout the pathways with cerebellar lesions predominating.

The diagrams illustrate most of the common causes of hydrocephalus, but there are some rare ones which are not included and two of them will be mentioned very briefly.

Rare varieties of hydrocephalus

The 'Kleeblattschädel' syndrome. One of the most striking manifestations of hydrocephalus is the Kleeblattschädel syndrome or the 'clover leaf' head which has been described in the German literature back as far as 1824.

The hydrocephalus is usually one part of the syndrome in which the ears point downwards often without external canals; there is proptosis with eroded corneae, and a recessed nose. In addition, there may be other anomalies, macrostomia, macroglossia, ankylosis of the elbows and knees. The honeycombed appearance of the squamous temporal bone on X-ray is quite classical. The condition is thought probably to be a variant of achondroplasia, which as is known is occasionally associated with hydrocephalus.

'Bobble-headed doll' syndrome. This very descriptive if unscientific title, used by Benton *et al* (1966) derives from an American term for those curious dolls seen in this country mainly in car rear windows, with heads on coiled springs so that they have a to-and-fro oscillation.

The syndrome consists of four features: hydrocephalus; the curious oscillations of the head, at the rate of about three per second; dysaesthesia of the head, so that patients dislike it being touched; and mental retardation. The syndrome is due to a congenital cyst, fibrous or hamartomatous arising from the third ventricle.

Fig. 3 shows a large cyst arising from the third ventricle in a patient of $5\frac{1}{2}$ years of age with the syndrome.

The clinical aspects of hydrocephalus form such a vast subject that I propose confining myself to two aspects: one is concerned with a recently developed method of investigation, and the other with one type of surgical treatment.

THE INVESTIGATION OF HYDROCEPHALUS

For localization of the obstruction in hydrocephalus air studies have become the standard method of investigation and have proved reasonably reliable. In communicating hydrocephalus the classical picture is that air enters the ventricles easily but does not progress above the interpeduncular or possibly suprasellar cistern. This has largely been confirmed by Granholm and Radberg (1963) who in thirty-seven congenital hydrocephalics demonstrated that no air progressed beyond the chiasma in thirty-three of them. However, in four cases, 11 per cent, some air was found to have made its way over the convexity. On the other hand they noted in another group of thirty-seven apparently normal infants without hydrocephalus that no less than fifteen, or 40 per cent, showed either no air at all or a very small quantity in the convexity subarachnoid space. This proportion is probably smaller in adults. Khibler, Couch and Crompton (1961) found in nearly every one of 150 normal cases air over the convexity, but did not give a quantitative assessment, and it is possible that in a number of patients the air may have been in very small amounts. However, it does appear that lack of convexity filling by air may occur in normal people, particularly children. In anatomical terms it may possibly represent a very shallow subarachnoid space with narrow communication with the major cisterns. Air studies, therefore, only give a partial, episodic and sometimes unreliable, picture of CSF circulation. The large amount that requires to be injected disturbs the hydrodynamics, as well as the patient, and there has been a need for a method complementary to air study to give details of the dynamics of CSF flow, such as speed and route taken using a material which would diffuse into, and outline all the subarachnoid space in which there was not a true mechanical obstruction. There would be many uses for such a tool; it would be helpful in diagnosis between communicating hydrocephalus and cerebral atrophy when one could eliminate the vagaries of air entry into the convexity subarachnoid space, which I referred to earlier, and draw reliable conclusions as to presence or absence of a block at the tentorial hiatus. It could also be used to demonstrate a blocked shunting system for which purpose air is unsatisfactory and in the presence of ventriculo-venous shunts possibly dangerous from air embolism. Again some cases of aqueduct stenosis and atresia of the exit foramina have obstructions in the subarachnoid space; that is to say a communicating as well as an obstructive hydrocephalus.

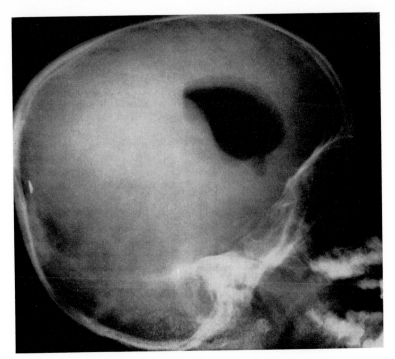

FIG. 3. Ventriculogram shows blockage of foramen of Monro by a large cyst which is bulging up into the lateral ventricle.

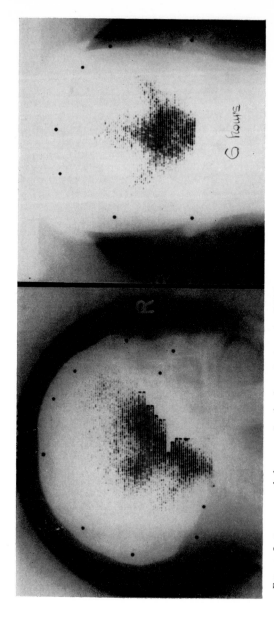

FIG. 4. Isotope encephalogram. In the lateral view the RIHSA is passing posteriorly into the ambient cistern and anteriorly into the suprasellar, chiasmatic and anterior end of callosal cistern. In the antero-posterior scan the callosal and Sylvian cisterns are filling. The movement of the isotope is delayed because the injection was made into the lumbar theca.

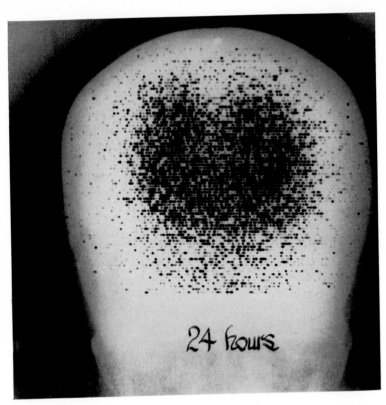

FIG. 5. Isotope encephalogram of a patient with long-standing communicating hydrocephalus.

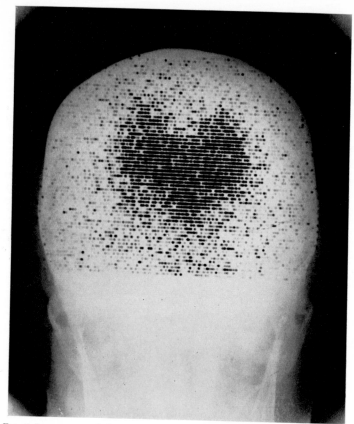

FIG. 6. Isotope encephalogram at 48 hours showing no movement of RIHSA out into the subarachnoid space in a patient with a blocked ventriculo-cisternostomy tube.

It would be helpful to know this prior to operation so as to avoid simply converting an obstructive into a communicating hydrocephalus requiring a second surgical procedure, and it is very difficult to reveal this reliably with air.

ISOTOPE ENCEPHALOGRAPHY

Recently human serum albumin labelled with ^{131}I (RIHSA) has been introduced for intrathecal scanning and as albumin is a normal constituent of CSF it was hoped that it would give some information regarding CSF flow.

The material is introduced as a very small injection $\frac{1}{2}$ to 1 ml, so that there is virtually no disturbance of CSF hydrodynamics and its gamma emission is then picked up by a conventional external scanner. The dosage is between 50 and 100 microcuries, apparently biologically unharmful to both central nervous system and the body generally. The injection is made into the cisterna magna rather than the lumbar theca because in the latter situation the RIHSA tends to be diluted too much, it takes too long to arrive in the head and only about a third of the total dosage gets there.

The flow pattern from the theca or cisterna magna in normal people is fairly consistent.

Within 45 minutes after a cisternal injection most of the isotope has flowed from the cisterna magna in front of the brain stem to the cisterna medularis, the cisterna pontis and then into the interpeduncular cistern, only a little passing down the spinal theca. From the cisterna interpeduncularis it moves along two routes in roughly equal proportions, backward through the paired ambient cisterns to the quadrigeminal cistern and posterior end of the callosal cistern. Anteriorly it passes from the suprasellar cistern laterally through the Sylvian and also forward through the chiasmatic cistern to that of the lamina terminalis and to the front end of the callosal cistern. All the supratentorial cisterns are well visualized at 4 hours (Fig. 4). From the callosal and Sylvian cisterns, isotope flows over the convexity towards the parasagittal area where at 12–24 hours post-injection most of the isotope is concentrated. By 48 hours it has cleared from the subarachnoid space virtually completely. There is no filling of ventricles.

In its passage the isotope will of course outline any tumours encroaching on the cisternae and any local obstruction by cysts or arachnoidal adhesions, but these lesions are probably better demonstrated by air and the main advantage of the method lies in evaluating hydrocephalus.

COMMUNICATING HYDROCEPHALUS

In communicating hydrocephalus the flow pattern is quite different to that in the normal person. The isotope collects entirely in the ventricles, which are by-passed in the normal, and may remain there for days. This pattern always indicates obstruction of the convexity subarachnoid pathways.

A patient who had had a large head from an early age became increasingly demented due to progressive communicating hydrocephalus. Fig. 5 shows the tracer collected in the ventricles and it is still there at 24 hours with no filling of the convexity subarachnoid space.

The use of the method in demonstrating blocked shunts is shown by the next case, a lady of 59, who in 1953 underwent a Torkildsen procedure, or ventriculo-cisternostomy for aqueduct stenosis. She remained well until 2 years ago, when she became slow, increasingly ataxic in walking and showed spasticity of both legs. A blocked Torkildsen's tube was thought possible and was investigated by an injection of RIHSA 100 microcuries directly into the ventricle. This scan at 48 hours (Fig. 6) shows no passage of isotope from the enlarged ventricles to the subarachnoid space, indicating a blocked tube.

The next application of the technique is the investigation of adult hydrocephalus with normal pressure, and the differential diagnosis from enlarged ventricles due to cerebral atrophy.

It has been taught in neurosurgical kindergarten for many years that in very large hydrocephalic ventricles resulting from obstruction of the third and fourth ventricles by lesions such as colloid cysts or aqueduct stenosis, the pressure as measured by ventriculography or spinal puncture may not be raised above normal, which I take to be 160 mm of water. However, by-passing of the block by shunting the large collection of fluid will often improve the neurological status of the patient. It was noted a long time ago, in 1956, by Foltz and Ward that in the communicating hydrocephalus following subarachnoid haemorrhage CSF pressure might also be normal but still associated with considerable mental retardation. We have been in the habit of draining externally the enlarged ventricles in patients with post-haemorrhagic hydrocephalus, and if their neurological status improved then a value shunt was inserted, often with considerable benefit irrespective of a high, normal or low pressure.

In the last 3 years attention has been drawn by Adams and his colleagues (1965) to the occurrence of hydrocephalus with pressures within the normal range which may occur apparently spontaneously, and this is the important feature, in middle-aged or elderly patients.

The case history is usually fairly simple, there is an initial story of headache suggestive of raised pressure and within a few months the symptomatology evolves over a period of weeks to several months and consists of impaired memory, slowness and poverty of thought, poor concentration, apathy and inertia, unsteadiness of gait and sometimes incontinence of urine.

Diagnosis has previously been made by air study which shows the typical block in the basal cisterns at the level usually, of the tentorium, the top of the interpeduncular cistern being bowed upwards. If these cases are drained with a low pressure valve down to say 60 mm of CSF, improvement occurs and is maintained if the shunt remains patent.

If diagnosed early, a complete return to normal may occur.

It seems there may be a number of elderly patients with this condition in whom it may not be diagnosed, the symptoms being attributed to pre-senile dementia or brain atrophy from arteriosclerosis and it is important to bear this fact in mind in differential diagnosis of dementias. The diagnosis of cerebral atrophy has previously depended on demonstrating air over the convexity subarachnoid space, but as mentioned earlier in some people little or no air goes over the surface despite the absence of obstruction and if this should occur in patients with cerebral atrophy a false picture of communicating hydrocephalus will be presented with the possibility of an unnecessary shunting procedure.

RIHSA is more reliable in giving an accurate picture and it causes less morbidity in these patients as Bannister, Gilford and Kocen (1967), in this hospital, have recently shown, and the characteristic pattern will be the collection of isotope in the ventricles; whereas in Alzheimer's disease and kindred dementias the isotope passes rapidly over the convexity without entering the ventricles and may in fact show an enhanced concentration in the parasagittal area.

THE TREATMENT OF HYDROCEPHALUS

The serious and routine surgical treatment of hydrocephalus, particularly the congenital variety, really started in 1957 with the introduction into surgery of the material silastic.

Although previously there were certain recognized procedures available, such as Torkildsen's procedure, ventriculo-peritoneal and spinoureteric diversions, the use of the valved ventriculo-venous shunt, made possible by silastic, meant that all forms of hydrocephalus could be treated with a low initial operative mortality in the region of 2–3 per cent, even in infants. Silastic causes virtually no reaction in human tissues and if a slit is made in its wall it will open at a predetermined pressure and close as soon as the pressure drops, and will do this almost indefinitely. This has been utilized to form a one-way slit valve permitting the passage downwards of the CSF but preventing the ingress of blood. At the present time there is a tendency to use valve systems to the exclusion of all other methods irrespective of whether the hydrocephalus is obstructive or communicating; however, the formidable list of complications of valve systems that has accrued over the years indicates that we have not yet achieved a perfect drainage system and a return to some of the older methods may be envisaged. With this in mind I will review the present procedures available in the treatment of hydrocephalus.

SHUNTING PROCEDURES
There are three main principles invoked in the surgical disposal of the CSF in hydrocephalus:

1   The reduction of the formation of CSF by removal of the choroidal plexus in each lateral ventricle sufficient for the fluid to be dealt with by what absorptive mechanisms still function—plexectomy.
2   The diversion of CSF past an obstruction and into the normal intra-cranial absorptive channels—a method which has the merit of recon-stituting the normal physiology but is applicable, of course, only to obstructive, or ventricular, hydrocephalus.
3   The re-routing of the CSF into the blood stream or other absorptive areas outside the central nervous system, a principle essential for the relief of communicating hydrocephalus but applicable also to the obstructive variety.

PLEXECTOMY
This method has its main advocate in Scarff (1966) in whose hands it seems to be satisfactory and consists of coagulating the choroid plexus of each lateral ventricle with a cystoscope-like instrument. It is a single treatment and no tubes are left behind to cause late complications. The reduction of CSF secretion is apparently adequate despite Bering's (1958) experimental findings in dogs in which removal of the plexus reduced the formation of CSF in the lateral ventricle by only 25 per cent. Scarff (1966) recently reviewed ninety-five cases treated in this way by three surgeons, himself, Feld and Putnam, with a high initial mortality, 15 per cent, a 65 per cent arrest rate and a low, 1 per cent, late complication rate.

The procedure has never attained much popularity. It may be that neurosurgeons generally have not acquired the necessary urological training.

Fig. 7 illustrates the routes taken by internal and external diversions. There are two internal diversions—Torkildesen's procedure and anterior ventriculostomy.

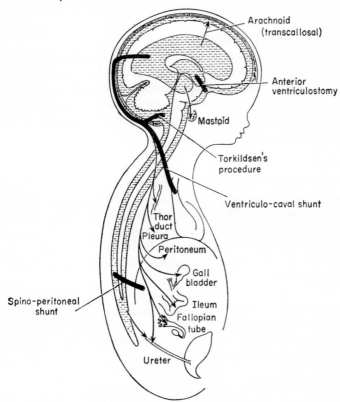

FIG. 7. To show the various diversionary procedures used for hydrocephalus. The black tubes illustrate procedures that are, or have been, extensively used.

TORKILDSEN'S PROCEDURE

Torkildsen's procedure, or ventriculo-cisternostomy, obviously can only be used for obstructive (ventricular) hydrocephalus, and once the tube has started working well, late complications are unusual. I have a number of cases in whom red rubber tubing was inserted, still functioning more than 16 years later. Modern silastic tubing should last almost for life. The intrinsic disadvantages of the method are that it is a major surgical procedure; it has a higher initial mortality rate than the venous shunts; the distal end of the cisterna magna is liable to be blocked by encroachment of the tonsils from continued tumour growth, which latter may also block the tentorial hiatus. In addition technical difficulties arise from the small posterior fossa and cisterna magna in comparison with the large head, particularly so in infants.

ANTERIOR VENTRICULOSTOMY

This method is not often used at the present time. I employed it a good deal in the past, inserting a slotted rigid polythene tube from the third ventricle to the chiasmatic cistern to prevent arachnoidal adhesions growing across the opening in the lamina terminalis. Once functioning, it tends not to block. It is technically quite a complicated procedure and unsuitable for young children.

EXTRA-CRANIAL DIVERSIONS

There are a number of extra-cranial shunting procedures and of these the valved ventriculo-venous shunt is the most popular for both communicating and obstructive hydrocephalus. It has the merit of being an external operation, it requires a relatively small incision in the neck and only a burrhole in the skull. There is no manipulation of the brain, nor gross changes in the CSF pressure at operation. This method will be referred to in detail later.

VENTRICULO-PLEURAL AND PERITONEAL SHUNTS

These are used sporadically, usually as an alternative when ventriculo-venous shunt has failed. They have an over-all success rate with revisions of 60–65 per cent, similar in fact to plexectomy.

ARACHNO- OR SPINO-PERITONEAL DIVERSIONS

This method is still in common use for communicating hydrocephalus, either as an initial treatment or an alternative to a blocked ventriculo-venous shunt. Again it has a 60–65 per cent success rate. Diversions into other target organs and channels—the mastoid, the ureter, gall bladder, thoracic duct and isolated portions of the ileum, have been illustrated for historical interest.

VENTRICULO-VENOUS DIVERSIONS

The universal application of the valved ventriculo-venous shunt to infants with either communicating or obstructive hydrocephalus has helped to lower considerably the initial mortality rate in treating con-genital hydrocephalus. The valves were introduced into England in 1958 so that we have had only 10 years' experience of them. There are two valve systems in common use, the Holter and the Pudenz-Heyer. The Hakim and the Till valves are employed less frequently for special indications. Use of the Holter and Pudenz valves varies according to individual surgical preference. Two small technical points are worth emphasizing: the ventricular catheter is inserted through an occipital burrhole along the whole length of the ventricle to the frontal horn, where there is no choroidal plexus to adhere to it, to block it or to bleed into it. The caval catheter is inserted via the jugular vein or a large branch so that the lower end reaches to the auriculo-caval junction. It is positioned under X-ray control, the tube being rendered visible by an injection of opaque contrast material so that the end can be seen to be at the correct situation on a level with the sixth thoracic vertebra. If too low it will enter the auricle and may cause contusion of its wall with the liberation of clots and predisposes to bacterial endocarditis. If too high in the superior vena cava where the blood flow is not sufficiently turbulent, the blood may clot in the tube or the vein itself may occasionally thombose.

The valves require to be revised prophylactically, usually in the second year of life and again between the fourth and sixth year in order to lengthen the caval catheter which, owing to the increasing discrepancy between the length of the catheter and the length of the child's chest, tends to be pulled higher up in the superior vena cava where the blood flow is less turbulent and clotting and obstructions of the vena cava may occur.

In addition to these physiological revisions there are the re-explorations demanded by obstruction or failure of the valve system. These compli-cations (Fig. 8) constitute a formidable list. This list relates particularly to hydrocephalus in infants and young children, of whom 70 per cent have, in addition, spina bifida with myelomeningocele. The present trend is to operate on meningoceles within 24 hours of birth, or less, to replace the neural tissue within the spinal canal and cover it up with healthy skin flaps. However, 75–80 per cent of these infants develop hydrocephalus in the next few weeks and 50 per cent of them will require a shunt procedure. These particular patients who tend to have high protein and often low-grade infection of the CSF, arising from the exposed neural plaque, both potent causes of shunt complications, comprise the majority of cases requiring revisions. In all infants the relatively narrow lumen of the superior vena cava will predispose to blocking of the lower catheter.

Q

Once a valve system has been inserted into the child the patient will require constant supervision thereafter.

Despite these complications the ventriculo-venous shunts still give better results in the young age group than any other procedure and provides a much better outlook than the limited treatment available up to 10 years ago.

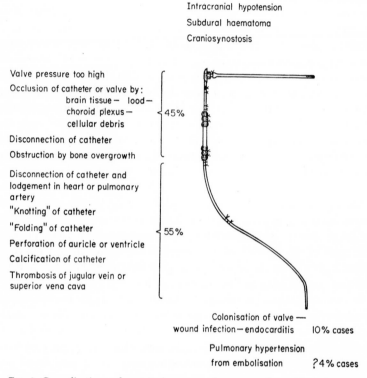

Intracranial hypotension
Subdural haematoma
Craniosynostosis

Valve pressure too high
Occlusion of catheter or valve by:
    brain tissue — lood—
    choroid plexus—   45%
    cellular debris
Disconnection of catheter
Obstruction by bone overgrowth

Disconnection of catheter and lodgement in heart or pulmonary artery
"Knotting" of catheter
"Folding" of catheter   55%
Perforation of auricle or ventricle
Calcification of catheter
Thrombosis of jugular vein or superior vena cava

Colonisation of valve —
wound infection—endocarditis   10% cases

Pulmonary hypertension
from embolisation   ?4% cases

FIG. 8. Complications of ventriculo-venous shunt procedures. Sixty per cent require early revision in infants and young children.

What happens when the child grows up? It is a daunting prospect to insert a valved shunt into a child of say 15 years of age with aqueduct stenosis with the implication, as the situation is at present, that he will have the tube poking into his right auricle for perhaps 60 years, and the pulmonary complications that may arise in such a case lend substance to this anxiety. This becomes even more frightening when a valve is inserted into a child of 15 months. It would be of interest to follow a group of infants through childhood and adolescence to maturity to ascertain the long-term outlook, to discover whether having reached adult life complications cease to occur and whether over a long period one of the other varieties of shunting procedure would prove less troublesome. As ventriculo-venous valve systems have only been employed for 10 years such an analysis is not possible. However, one could probably draw some inferences from the behaviour of these shunting procedures when carried out primarily on adults.

TABLE I

Aetiology of hydrocephalus in cases treated by ventriculo-venous diversions

Obstructive hydrocephalus (*22 cases*) *Aetiology*		*Communicating hydrocephalus* (*14 cases*) *Aetiology*	
Aqueduct stenosis	6	Unknown	3
Pineal tumour	4	Subarachnoid haemorrhage	3
Brain-stem angioma	1	Cysticercosis	1
Brain-stem glioma	1	T.B. meningitis	1
Cerebellar astrocytoma	1	Sarcoid	1
Cerebellar ectopia	1	Tuberose sclerosis	1
Dandy-Walker	1	C.C.I.	1
Cysticercosis	1	Post-operative	3
Third ventricle tumour	1	Frontal meningioma	
Thalamic tumour	1	Cervical neurofibroma	
Craniopharyngioma	1	Cholesteatoma	
Intracranial abscess	1		
Bifrontal astrocytoma	1		
Cholesteatoma	1		

Up to the present time there has been very little information about the efficacy of ventriculo-venous valves in adults. Theoretically the valves should be relatively trouble free in comparison with younger people, the superior vena cava has a much larger lumen so is less liable to thrombose, the blood flow is more turbulent and so there is less liability to clotting in the tube; there is no problem with chest elongation. On the other hand the ventricles in adult hydrocephalus do not attain the enormous size of the congenital forms so the obstruction rate of the catheter in the ventricle may be increased. To assess these various points all the ventriculo-venous

TABLE 2

Aetiology of obstructive hydrocephalus in cases treated by Torkildsen's
procedure (ventriculo-cisternostomy)

Torkildsen procedure (45 cases)	
Aqueduct stenosis	10
Third ventricular glioma	7
Thalamic tumour	6
Pineal and tectal plate tumours	5
Colloid cyst	3
Brain-stem glioma	3
Craniopharyngioma	2
Frontal and septal tumours	2
Fourth ventricle tumour	2
Pituitary tumours	2
Post-meningitic	2
Cerebellar tumour	1

TABLE 3

Results of ventriculo-venous shunts. The success rate for obstructive hydrocephalus comprises fourteen patients still alive with functioning valve systems and two patients who died from tumour growth but whose valve systems remained patent

Ventriculo-venous shunts (36 cases)					
		Revisions	*Shunt patent*	*Shunt blocked*	*Purpose of shunt fulfilled*
Obstructive hydrocephalus	Alive 14	6 cases (10 procedures)	14		16 = 73%
	Dead 8		2	6	
					29 = 80%
Communicating hydrocephalus	Alive 13	6 cases (8 procedures) (1 removal)	13		13 = 93%
	Dead 1	—		1	

shunts and all the Torkildsen's procedures which have been carried out on patients over the age of 16 in my department, have been analysed. The methods used for both procedures are standardized so the results, therefore, are reasonably comparable. Tables 1 and 2 illustrate the various indications for the two types of procedures.

In the obstructive group of those cases treated by ventriculo-venous shunt the proportion of benign lesions, such as aqueduct stenosis, Dandy-Walker syndrome, etc., stand at 40 per cent, whereas they comprise only 35 per cent of the Torkildsen-treated cases, so that the former group as a whole has a slightly more favourable prognosis.

RESULTS

In assessing the results, those cases who died but nevertheless whose valves or tubes had remained patent to the end, so that the purpose of the shunt was fulfilled, have been included in the total success rate for each procedure (Table 3).

With regard to ventriculo-venous shunts the postulates made with regard to complications are not far wrong. Out of the eighteen revisions not one was for the caval catheter block. All were for the blocks in the ventricular catheter or valve itself. However, the 50 per cent revision rate is only a little lower than the 60 per cent in children, so the method is not trouble free by any means in adults. The two cases dying with the shunt patent did so as a result of continued tumour growth, so that the purpose of the shunt in relieving the hydrocephalus was successfully achieved.

At first sight the results, with revision, in adults appear reasonably good with 93 per cent success rate for communicating hydrocephalus and 80 per cent for communicating and obstructive hydrocephalus combined. However, the length of follow-up is only 2 years 2 months and it is reasonable to expect that the success rate, at least for obstructive hydrocephalus, will fall further with the passage of time. If the results of ventriculo-venous shunts for obstructive hydrocephalics are analysed separately, it will be seen that of the twenty-two cases, six have already blocked in a follow-up of just over 2 years.

It is reassuring to observe the good results, over 90 per cent success rate, in communicating hydrocephalus, so the prospects for children with this condition who attain adult life are promising.

Table 4 shows the results of the Torkildsen's procedure, from which it will be noted that a 78 per cent success rate pertains after a follow-up of an average of over 5 years. Of the twenty-four deaths, fourteen patients

TABLE 4

The results of the Torkildsen procedure show a 78 per cent success
rate which comprises twenty-one patients alive, and fourteen who
died from tumour extension but whose ventriculo-cisternostomy
tube remained patent throught

Torkildsen procedure (45 cases)

	Tube patent	Tube blocked	Purpose of shunt fulfilled	Duration of follow-up
Alive 21	21	—		Av. 6 years (3 days—15 years)
			35 = 78%	
Dead 24	14			Av. 4 years 2 months
		10		(2 years—16 years)

TABLE 5

Torkildsen procedures compared with ventriculo-venous shunts performed
in the treatment of obstructive hydrocephalus

Torkildsen procedure

		Tube patent	Tube blocked	Purpose of shunt fulfilled	Duration of follow-up
	Alive 21	21	—		
45 cases				35 = 78%	Av. 5 years 3 months (3 days—16 years)
	Dead 24	14			
			10		

Ventriculo-venous shunts

		Shunt patent	Shunt blocked	Purpose of shunt fulfilled	Duration of follow-up
	Alive 14	14	—		
22 cases				16 = 73%	Av. 2 years 2 months (4 months—7 years)
	Dead 8	2			
			6		

died from continued tumour growth but with the tubes still patent, giving a 78 per cent success rate in which the purpose of the shunt was fulfilled. The over-all success rate for both Torkildsen and ventriculo-venous shunts is about 80 per cent. However, when comparing ventriculo-venous shunts for obstructive hydrocephalus (Table 5) it will be seen that the figure has fallen to 73 per cent, after an average follow-up of only 2 years 2 months, and with the passage of time the results must fall lower still. Selection of cases does not play a part in these results for, as mentioned earlier, the proportion of neoplastic to benign obstructions is biased against the Torkildsen cases.

It would seem that ventriculo-cisternostomy for obstructive hydro-cephalus in adults has better results than ventriculo-venous shunts. The reason probably resides in the fact that larger drainage tubes are used with less liability to block with the protein leak from the tumours and the ventricles do not collapse down to the same extent as they do when the CSF is diverted into the vena cava.

The practical conclusion to be drawn from these figures is that in adults with obstructive hydrocephalus from a tumour, with a short life expectation of a few months, one would use a ventriculo-venous shunt because of the simplicity of its insertion and the fact that it has a reasonable patency rate for a short time. However, with obstructions associated with a longer life expectation, particularly for aqueduct stenosis, the Torkild-sen's procedure would appear to be more reliable and my practice now is to use ventriculo-cisternostomy in the first instance and to have recourse to ventriculo-venous shunts only in the event of its failure.

To sum up the methods of treatment in general:

In childhood for both communicating and obstructive hydrocephalus ventriculo-venous shunts carry the best prognosis despite the formidable complication rate, which is partly compensated for by the low initial mortality.

In adults with communicating hydrocephalus the valved shunt works reasonably well with a high revision rate, in particular there seem to be few complications from the caval catheter, but only the passage of many years will show whether pulmonary embolization is going to be an important feature. In adults with obstructive hydrocephalus, particularly with a long life expectation, the Torkildsen procedure may well prove the better with a ventriculo-venous shunt held in reserve in case of failure.

This is how the surgical treatment of hydrocephalus rests at the present time, awaiting development of biochemical methods designed to control the secretion and absorption of cerebrospinal fluid.

ACKNOWLEDGEMENTS

I am grateful to Dr J.W.D.Bull and Dr D.Sutton and their respective radiological departments for the radiographs and the reproductions of the isotope encephalograms, and also to Mr Wylie McKissock for the case history of the patient whose gamma scan is illustrated in Fig. 6.

REFERENCES

ADAMS R.D., FISHER C.M., HAKIM S., OJEMANN R.D. and SWEET W.H. (1963) *New Engl. J. Med.* **273,** 117

BANNISTER R., GILFORD E. and KOCEN R. (1966) *Lancet* **ii,** 1014

BENTON J.W., NELLHAUS G., HUTTENLOCHER P.R., OJEMANN R.D. and DODGE P.R. (1966) *Neurology (Minneap.)* **16,** 725

BERING E.A. (1958) *Clin. Neurosurg.* **5,** 77

DANDY W. and BLACKFAN K.D. (1914) *Am. J. Dis. Child.* **8,** 406

FOLTZ E.L. and WARD A.A. (1956) *J. Neurosurg.* **13,** 546

GRANHOLM L. and RADBERG C. (1963) *J. Neurosurg.* **20,** 338

KHIBLER R.F., COUCH R.S.C. and CROMPTON M.R. (1961) *Brain* **84,** 62

LUESSENHOP A.J. (1966) *Clin. Proc. Child. Hosp. (Wash.)* **22,** 112

RAY, B.S. and PECK F.C. (1956) *J. Neurosurg.* **13,** 317

SCARFF J.E. (1966) *Archs Neurol. Psychiatry* **14,** 382

WELCH K. and FRIEDMAN V. (1960) *Brain* **83,** 454

CHAPTER SIXTEEN

THE CEREBROSPINAL FLUID PRESSURE

HUGH DAVSON

In the following pages I shall discuss the cerebrospinal fluid (CSF) with special reference to the pressure within the cerebrospinal system, i.e. the pressure above or below atmospheric that is registered when a cannula is inserted into the fluid and connected to a suitable pressure-transducer. When measured at the lumbar sac in man in the lateral recumbent position the pressure is normally positive, being on average some 150 mm H_2O; in the dog, measured at the cisterna magna in the recumbent position, the pressure is likewise positive; thus Becht found an average value of 112 mm H_2O with extremes of 242 and 43 mm H_2O.

A. PRESSURE-FLOW RELATIONSHIP

The cerebrospinal fluid is being formed continuously and this continual production of fluid must be a factor in determining the steady level of the CSF pressure. Thus there is every reason to believe that simple laws of fluid motion should apply, so that the rate of flow through the system will be given by the pressure-head and the resistance to flow. The pressure-head in these circumstances will be the difference between the CSF pressure and the pressure in the blood vessel into which the fluid drains, namely the dural sinus pressure. Thus:

$$\text{Flow} = P_{CSF} - P_{Sinus/Resistance}$$

The resistance will be the simple Poiseuille-type viscous resistance to flow through the subarachnoid spaces, and probably the most significant site of this resistance is normally the arachnoid villi, spongy erosions of the dura that give access from subarachnoid space to the lumen of the dural sinus and lacunae laterales.

STEADY STATE

Before passing to a consideration of the factors determining the flow of cerebrospinal fluid, it must be emphasized that the simple equation above applies only to a steady-state condition which may well be an ideal rather than a practical reality. Thus a sudden change in posture will make a profound change in the recorded pressure and this will be mainly due to a change in the amounts of blood in the large veins of the head and cord; again, alterations in vascular pressures, especially pressures in the large veins, may have large effects on CSF pressure which need not be connected at all with changes in flow-rate of cerebrospinal fluid or with changes in resistance. It must be appreciated, then, that the CSF pressure may be very strongly influenced by vascular pressures, so that alterations in these may for the time being completely dominate the actual pressure. Only after the system has settled to a steady state will the simple equation be applicable. Thus, to take a simple example, we may transfer a dog from a horizontal position to the head-down position; because of the engorgement of the cerebral veins the CSF pressure rises rapidly, but if the new posture is maintained indefinitely there is a gradual return to a lower CSF pressure because the raised pressure causes a greater flow of cerebrospinal fluid out of the cranial cavity and thus allows some of the extra blood to be accommodated. The pressure may be expected to remain higher in this new posture because of the greater column of fluid whose weight comes into play, but finally a new steady state is established with the simple flow-equation applying, although the actual values of some of the parameters may be different; thus it is likely that flow-rate, which is determined by rate of secretion, will remain the same, but the CSF pressure and sinus pressure may both be raised; if the resistance is unaltered, then the pressure-head, namely $P_{CSF} - P_{Sinus}$ will have returned to its original value.

The investigation of the physiological mechanisms concerned in maintaining the normal CSF pressure, and the raised pressures in hydrocephalus, must be concerned with all the separate parameters of the simple equation, namely the rate of secretion, the resistance to flow and the pressure in the dural sinuses.

B. PRODUCTION OF THE CEREBROSPINAL FLUID

Since Dandy's classical studies, it has been generally accepted that the choroid plexuses are the main, if not the exclusive, source of the cerebrospinal fluid. These structures, which are simple outpouchings of the pia into the ventricular cavities, are probably ideal structures for the elaboration of a fluid secretion; being highly vascular they are able to produce, within their stroma, an ultrafiltrate of plasma; whilst the specialized epithelial layer interposed between the stroma and the ventricular cavity —the *choroidal epithelium*—has many of the features that we have come to regard as characteristic of a fluid-secreting system; thus, in the electron microscope, the epithelial cells exhibit a series of basal infoldings of their membranes that are also seen in the ciliary epithelium of the eye, in the epithelium of the proximal tubule and in the salt-gland of birds, all epithelia that are noted for continuous transport of water. Thus, on morphological grounds, we would be justified in supposing that the choroid plexuses were the site of secretion of the cerebrospinal fluid. The classical experiments of Dandy and Blackfan (1914) and Frazier and Peet (1914), in which an experimental hydrocephalus was produced in dogs by plugging the aqueduct of Sylvius, provided strong evidence for a ventricular source of the cerebrospinal fluid; moreover, when the plexus of one lateral ventricle was removed, under these conditions, and both foramina of Monro were blocked, the ventricle containing the plexus expanded whilst the other contracted in volume (Dandy, 1919). As Bering (1962) has pointed out, however, the interpretation of this experiment is not unequivocal since the assumption implicit in the experiment is that, if the foramina of Monro had been left open, the hydrocephalus would have been symmetrical, the fluid from the normal ventricle forcing itself into the plexectomized one. In fact, when Bering repeated the experiment leaving both foramina open, the hydrocephalus was still asymmetrical. Bering argued strongly in favour of a non-ventricular source of fluid, or at any rate a non-plexus source, the fluid being formed presumably by the nervous tissue and passing from there into the subarachnoid spaces; certainly Bering & Sato (1963) have provided some evidence, based on experiments on dogs made hydrocephalic by blockage of the normal drainage routes, favouring a production of fluid directly into the subarachnoid spaces. Again, Pollay and Curl (1967) has perfused the aqueduct of Sylvius and obtained evidence for production of fluid in this channel that could not have come from any of the plexuses.

Positive evidence in favour of secretion by the choroid plexuses was provided by Welch (1963) who inserted a cannula into the large choroidal vein that drains most of the blood from the lateral ventricle choroid plexus in the rabbit; the emerging blood was shown to have lost some of its water, since the haematocrit, expressed as the proportion of cells to whole blood, was greater in this emerging blood than in the arterial blood. From rather crude estimates of the rate of blood flow through the plexus, Welch computed that the total production of fluid by all plexuses in the rabbit would be some 8 μl/min. The value for the rabbit given by Bradbury and Davson (1964) was 10.1 μl/min; this is based on inulin dilution and may be a high estimate since there is some loss of inulin into the brain (Segal and Davson, 1968). Thus, if Welch's estimate is correct, there is not much room for a non-plexus supply of fluid.

EXTRACELLULAR FLUID

If there is, indeed, a non-plexus site of formation of the fluid, we must ask what cells are responsible for the production. In this connection we must appreciate that the extracellular spaces of the central nervous tissue contain a fluid that may be regarded as a secretion very similar in composition to the cerebrospinal fluid. This fluid could be completely stagnant, but it seems more likely that it is formed continuously, even if very slowly, and if so it would presumably have to drain into the ventricles and subarachnoid spaces. The cells responsible for the formation of this extracellular fluid could either be the endothelial cells of the capillary wall or, more likely, the astrocytes whose end-feet make such close relationship with the basement membrane of the endothelial cell. These cells could remove the constituents of blood plasma, less its proteins, from the capillary and elaborate a secretion that is transported outwards into the intercellular space of the tissue. The evidence favouring a secreted extracellular fluid, similar in composition to the cerebrospinal fluid, is indirect and based on studies of exchanges of ions and other solutes between cerebrospinal fluid and the nervous tissue (see, for example, Davson, 1958; Davson and Bradbury, 1964; Pappenheimer, Fencl, Heisey and Held, 1965). The presence of a gentle current of fluid out of the tissue into the cerebrospinal cavities—ventricles and subarachnoid spaces—would emphasize the lymphatic role of the cerebrospinal fluid, in the sense that it carried away unwanted products of metabolism and cellular breakdown that were unable to pass easily into the capillaries of the tissue.

RATE OF PRODUCTION OF FLUID

The study of the factors determining the production of the cerebrospinal fluid is greatly facilitated by the development of an accurate method of estimating the rate of secretion of the fluid. In experimental animals this can now be done with some precision; thus a substance that is unlikely to cross the ependymal and pial walls of the cavities may be injected into a ventricle and the rate of its appearance in the blood or urine may be determined. A useful marker for this sort of study is labelled protein (Sweet, Selverstone, Solloway and Stetten, 1950; Van Wart, Dupont and Kraintz, 1960) or inulin (Rothman *et al*, 1961). A more precise method consists in perfusing the ventricular system through cannulae inserted in the lateral ventricles, drainage of the perfusate being brought about through another cannula in the cisterna magna. If the perfusion fluid contains a marker that does not escape significantly into the adjacent tissue, then the change in concentration that it undergoes will be a measure of the addition of freshly secreted fluid to the perfusate (Pappenheimer

TABLE I

Rates of secretion of cerebrospinal fluid

Species	$\mu l/min$	Per cent/ min	$\mu l/min/mg$ c.p.
Rat	2.2	—	—
Rabbit	10.1	0.63	0.43
Cat	20	0.45	0.5
	22	0.50	0.55
	22	0.50	0.55
Dog	50	0.40	0.625
	66	—	0.77
	47	—	—
Goat	164	0.65	0.36
Man	520	0.37	0.29

et al, 1962). Table I gives estimates of the rate of secretion, based on this technique, in a number of species; although the actual rates vary greatly, it will be seen that the fraction of the total volume produced per minute is much less variable and is in the region of 0.4 per cent. It seems that the turnover rate, expressed in this way, increases a little with decreasing size of animal.

FACTORS AFFECTING RATE OF SECRETION
It is known that in obstructive conditions very high fluid pressures may develop; it is reasonable to conclude, therefore, that the secretory mechanisms are able to operate in the face of these high pressures, in a similar way to the secretion of saliva that may take place against a pressure equal to the arterial pressure. Evidence on this point is not extensive; Heisey, Held and Pappenheimer (1962) found no significant change in rate of

secretion in the goat when the fluid pressure was varied between +28 and − 10 cm H_2O; on the other hand Flexner (1933) claimed that, in the cat, the rate of secretion varied inversely with the pressure; he used Flexner and Winters's (1932) technique of inserting a cannula into the aqueduct of Sylvius, leakage around the cannula being prevented by a rubber balloon. The secreted cerebrospinal fluid was allowed to run into a graduated horizontal tube, and its rate of flow measured when different pressures were established to oppose outflow. As indicated above, they found the rate at which fluid passed into the calibrated tube to decrease with increasing applied pressure. Unfortunately the interpretation is not unequivocal; the high pressure of fluid within the ventricles might well cause a slow passage of fluid into the tissue of the brain and this would be reflected in a diminished outflow into the calibrated tube. If the high pressure caused some ischaemia of the choroid plexuses, then it is reasonable to believe that the rate of secretion of fluid would be reduced; thus noradrenaline perfused through the ventricles caused a reduction in rate of production of fluid by some 20–40 per cent (Vates, Bonting and Oppelt, 1964). At any rate, until more work is done on this point, we must maintain an open mind as to whether the rate of secretion of fluid is significantly diminished when the pressure rises as a result of obstructions, as in hydrocephalus.

INHIBITION

The rate of secretion of cerebrospinal fluid may be dramatically reduced by administration of a carbonic anhydrase inhibitor, such as acetazoleamide or Diamox (Tschirgi, Frost and Taylor, 1954; Davson and Luck, 1957; Fishman, 1959; Van Wart, Dupont and Kraintz, 1961; Davson and Pollay, 1963; Oppelt, Patlak and Rall, 1964). The reduction in the secretion rate, determined in the intact animal, is of the order of 50 per cent. The mechanism by which carbonic anhydrase inhibition brings about this reduction in rate of secretion is by no means clear; in other systems it has a similar effect, e.g. the aqueous humour of the eye, the salt-gland of certain birds, the gall bladder and so on. It was argued by Tschirgi, Frost and Taylor that the prime step in the secretion of the cerebrospinal fluid was the conversion of CO_2, formed in the brain, to bicarbonate; thus these authors were inclined to see the source of cerebrospinal fluid in the brain tissue, rather than the choroid plexuses. If this were true, we should expect the drug to inhibit the turnover of [24]Na in the brain in the same way that it inhibits it in the cerebrospinal fluid; but Davson and Luck (1957) found no change in this turnover, and it would seem that the small effects described by Koch and Woodbury (1960) in the rat are secondary to the primary change in turnover in the cerebrospinal fluid (Segal and Davson, 1958).

Cardiac glycosides

It would seem that all active transport processes in which transport of Na^+ is concerned, either directly or indirectly, are strongly inhibited by cardiac glycosides. These act by inhibiting a specific enzyme—the so-called *Na-K-activated ATPase*—that breaks down ATP and makes the energy so liberated available for the active transport process (Skou, 1957, 1960; Post and Jolly, 1957). In the cat, Vates *et al* (1964) reduced the rate of secretion to some 26 per cent of the control value when 1.10^{-4} M ouabain was incorporated into the medium during a ventriculo-cisternal perfusion. At this concentration, however, there is a great risk of killing the animal, and it would seem that, in contrast with Diamox, the cardiac glycosides are unlikely to be therapeutically useful in the control of the production of the cerebrospinal fluid. A similar situation pertains in the eye; locally injected ouabain will reduce formation of aqueous humour, but attempts to influence production by systemic administration have failed. This is understandable because active transport of Na^+ is vitally concerned in neural processes, so that a blanket inhibition of this through-out the body must have fatal consequences; by contrast, inhibition of carbonic anhydrase merely slows down a process that can occur in the absence of the enzyme.

Hypothermia

Experimentally, a considerable reduction in rate of secretion of cere-brospinal fluid may be brought about by a hypothermia of some 16° C; this is reflected in a reduction in the rate of turnover of ^{24}Na in the cere-brospinal fluid (Davson and Spaziani, 1962). It seems likely that the effects are due to an ischaemia of the choroid plexuses rather than to a simple inhibition of secretion by the low temperature *per se;* at any rate, the rate of secretion may be restricted to about that of normothermic animals by taking steps to return the systemic arterial pressure to the normal range.

OVERPRODUCTION OF FLUID

The clinician concerned with hydrocephalus will naturally ask whether the condition could arise through a pathological increase in rate of production of cerebrospinal fluid, as well as through an obstruction to the outflow routes. Physiologically an increased rate of production of fluid, due to the exhibition of some drug, has not been described. The pathologically raised CSF pressure in vitamin-A deficiency, where there is often no evidence for obstruction to outflow, has suggested an increased rate of production of fluid as the primary aetiological factor (Woollam and Millen, 1958), but no histological evidence implicating the arachnoid villi has been presented, and the suggestion of Mellanby (1941) that the obstruction is due to bone malformations, leading to obstruction of the aqueduct of Sylvius, is unlikely to provide a general explanation, since raised pressure occurs in vitamin-A-deficient animals without evidence of bone malformations (Moore and Sykes, 1941). In this connection we must note that, in order to double the CSF pressure, the rate of production of fluid, other things being equal, must be more than doubled, since examination of the simple equation relating flow to pressure and resistance shows that it is the *difference of pressure*, $P_{CSF} - P_{Sinus}$ that varies linearly with flow (F). Thus, if the value of P_{CSF} is 150 mm H_2O and the value of P_{Sinus} is 90 mm H_2O, the pressure difference is 60 mm H_2O. If F is doubled, and the resistance remains the same, the pressure difference will double, so that $P_{CSF} - P_{Sinus}$ becomes 120 mm H_2O. The Sinus pressure will be unaffected by the increased rate of flow, so that P_{CSF} must become 210 mm H_2O to give the doubled pressure difference:

$$P_{CSF} - P_{Sinus} = \text{Pressure difference}$$
$$210 - 90 = 120.$$

Thus doubling the rate of flow has only increased the CSF pressure by $60/150 \times 100 = 40$ per cent. To double the CSF pressure, i.e. to make it 300 mm H_2O, we must increase the rate of flow to 350 per cent of its original value. On these grounds, then, it seems unlikely that very large increases in CSF pressure will be caused by an increased rate of production of fluid alone.

C. THE DRAINAGE MECHANISM

The flow of fluid out of the cerebrospinal system will be determined by two factors, namely resistance to flow through the drainage channels and the pressure in the blood vessels into which these channels open. These factors are discussed by Logan, and here it is sufficient to indicate the necessity for a modification of Weed's concept of the mechanism of transport of fluid from the subarachnoid space into the dural sinus. According to Weed (1935) this occurred by virtue of the difference of colloid osmotic pressure between plasma and cerebrospinal fluid, similar to the absorption of extracellular fluid by a blood capillary. The difference of colloid osmotic pressure is large by virtue of the low concentration of protein in the cerebrospinal fluid, and this, combined with the hydrostatic difference of pressure, given by $P_{CSF} - P_{Sinus}$, would provide a powerful absorptive tendency. Experimentally Weed (1935) supported this concept by replacing the cerebrospinal fluid with artificial fluids containing larger and larger concentrations of protein and showing that, as the concentration increased, the absorption of the fluid, as measured by the tendency for fluid to flow into the cisterna magna through an inserted cannula, tended to decrease with increasing concentration. As pointed out earlier (Davson, 1956) Weed was probably measuring the tendency for the artificial cerebrospinal fluid to absorb fluid from the brain, a tendency that would increase as he increased the concentration of colloidal material in the fluid, and a tendency that would be reflected in a diminished flow into the system through the cannula. On theoretical grounds, moreover, Weed's system is unsound; thus if the cerebrospinal fluid is to be able to carry away with it the proteins that it contains, it must pass through a channel that places little or no restraint on these large molecules; such a channel, however, would not permit the development of a difference of osmotic pressure due to a difference in concentration of protein. Thus Weed's system, because of the postulated impermeability to protein of the drainage channel, would prevent the clearance of proteins from the cerebrospinal system. The continuous production of cerebrospinal fluid with some protein in it, the addition of further protein by diffusion from the brain, and the drainage by channels that caused the proteins to be retained would lead, ultimately, to the protein concentration in cerebrospinal fluid rising to that in the plasma and thus to a cessation of the colloid-osmotic absorption. Thus the channels permitting the passage of cerebrospinal fluid into the blood must be freely permeable to proteins, and it is of interest that Welch and Friedman (1960) have demonstrated both histologically and physiologically that the channels are indeed so large as to place no restraint on the passage of the largest protein molecules. The valvular nature of these openings prevents a reflux of blood from dural sinus into the subarachnoid space and, moreover, permits the development of a positive CSF pressure; thus only when the pressure built up to some 10 mm H_2O did the valves open to permit drainage.

OBSTRUCTED DRAINAGE

So far as obstruction to drainage is concerned, the physiologist has little to offer in the way of experimental facts that throw any light on the subject. The spongy nature of the arachnoid villi, through the interstices of

R

which flow of fluid must occur, suggests that occlusion of this pathway is a real possibility and may well be a factor in communicating hydrocephalus. In experimental animals, moreover, hydrocephalus may be induced by injection of particulate suspensions into the subarachnoid space; and this presumably is brought about by mechanical clogging of the pores in the villi. In the dog, however, a single injection seems not to be adequate, because there are several drainage routes. Thus, the main outflow is probably through the arachnoid villi into the dural sinuses; and after a single injection of particulate matter the strong flow along this route means that the arachnoid villi become blocked; subsidiary drainage channels along the optic and olfactory nerves, however, are spared, so that hydrocephalus does not develop, and only after a second injection do these subsidiary channels become blocked (Schurr, McLaurin and Ingraham, 1953). In the cat neither of these subsidiary channels is significant, and in the monkey only that along the optic nerves.

SPINAL FLUID

A problem that has occupied the experimentalist frequently is the behaviour of the fluid in the spinal subarachnoid space; chemically speaking the fluid here appears to be far less exposed to change than the ventricular fluid, and on this account the fluid might be considered to be in a backwater from the main current of fluid. Thus when ^{24}Na is injected into the blood, it appears rapidly in the ventricles and cisterna magna, but the turnover in the spinal fluid is very much slower (Sweet *et al*, 1950; Tubiana, Benda and Constans, 1951). Some mixing of spinal and cranial fluids does occur, however; this is well demonstrated by modern isotope scanning techniques in which a human subject is given a lumbar injection of radioactive colloidal gold and the activity in the brain and cord is scanned by an appropriate external counter. With passage of time the radioactivity reaches the basal cisterns and ultimately the subarachnoid spaces of the convexity (Rieselbach, Di Chiro, Freireich and Rall, 1962). Since mixing of spinal with cranial fluid does occur, the question arises as to whether there are consistent currents that bring about this mixing, or whether this occurs as a result of adventitious alterations in the relative pressures in the cranial and spinal subarachnoid spaces. A consistent current could be brought about by a drainage mechanism in the lumbar region, and the presence of arachnoid granulations similar to the arachnoid villi of the cranial dural sinuses have been described in the spinal venous plexuses; like the arachnoid villi they are ingrowths of arachnoid tissue into the wall of the blood vessel (Elman, 1923; Welch and Pollay, 1963).

However, experiments on the completely immobilized animal have shown that there are negligible currents in the spinal subarachnoid space; shaking the animal, or inducing convulsions, causes relatively rapid movements of fluid, so that we must conclude that mixing of spinal fluid with the cranial fluid, and consequently the renewal of the spinal fluid, depends on changes in relative pressure in the two compartments; these changes in pressure may well be induced by changes in venous pressure in the spinal column and cranium, pressures that are peculiarly susceptible to alterations in posture (Grundy, 1962).

Spinal cord
The brain has an internal cerebrospinal fluid in the ventricles and an external fluid in the subarachnoid spaces; the spinal cord has a large subarachnoid space but the internal fluid is very small in amount, being contained in the spinal canal which is a mere vestige of the embryonic neural tube. A flow down this canal into the sacral subarachnoid space, if it occurred, would be a means of circulating the spinal subarachnoid fluid. In the rabbit Bradbury and Lathem (1965) observed a definite flow of fluid down this canal, as manifest in the passage of a dye within it after injection into a lateral ventricle. Later, the dye appeared in the lumbar subarachnoid space, so that the picture after a time was one of the canal being strongly coloured along its whole length whilst the caudal and sacral subarachnoid spaces were also coloured with an intervening colourless region of the subarachnoid space. Thus, the rapid appearance of the dye in the sacral subarachnoid space could not have been due to flow from the cisterna magna down the spinal subarachnoid space. Histological examination showed that the dye escaped from the spinal canal by way of the filum terminale. In all the other species examined, namely the rhesus monkey, cat, rat, guinea-pig and sea lamprey, no evidence of flow along the canal could be obtained.

D. HYPEROSMOLAL SOLUTIONS

Withdrawal of cerebrospinal fluid from the subarachnoid space, e.g. from the human lumbar sac, will cause a profound fall in CSF pressure for obvious reasons; the pressure will return slowly to normal as the lost fluid is replaced; and this recovery will be facilitated by the reduction or complete cessation of drainage by virtue of the reduced CSF pressure. As a treatment for raised CSF pressure, then, tapping of the fluid can only be palliative. The same effect can be obtained by osmostic means; if the subject is given an intravenous injection of hypertonic urea there is a rapid fall in CSF pressure (Smythe, Smythe and Settlage, 1950), which is mainly due to an actual shrinkage of the brain due to the circumstance that urea does not cross the blood-brain barrier easily, whereas water does. The hyperosmolality of the blood, induced by the urea injection, thus causes passage of water from the extracellular fluid of brain into the blood. The extracellular fluid then becomes hyperosmolal to the cells of the brain and these lose water which passes out to the blood. The fall in CSF pressure brings drainage of cerebrospinal fluid to a stop, and part of the extra space made available in the cranial cavity by the shrinkage of the brain is actually made up by newly formed cerebrospinal fluid (Rosomoff, 1961, 1962). The maximal effects are observed in about 30 minutes, and within 6 hours the relative volumes of brain, cerebrospinal fluid and blood within the cranial cavity have returned to their original values. If Diamox is given as well as hyperosmolal urea the fall in CSF pressure is more sustained because the rate of formation of cerebrospinal fluid is reduced (Reed and Woodbury, 1962). Clearly, any substance that does not easily penetrate the blood-brain barrier will be effective, and the main virtue of urea in this respect is the tolerance of human subjects to large quantities administered intravenously (about 1–1.5 g per kg body-wt) and to its relatively slow elimination in the urine. Glycerol is effective and may be taken orally, but its renal elimination is more rapid than that of urea.

REBOUND

A possible danger in the use of hyperosmolal urea for the acute lowering of CSF pressure is the rebound in pressure that has been described following the return to the original value. Theoretically this may be expected, since urea does penetrate into the brain, albeit slowly, so that when, as a result of renal excretion, the blood level has fallen back to its original value, the concentration of urea in brain may well be above normal and thus favour passage of water from blood to brain. The balance of recent evidence suggests, however, that the rebound is an artefact (Javid, Gilboe and Cesario, 1964), and certainly any reversal in osmotic relations between blood and brain is not likely to be large (Pappius and Dayes, 1965).

E. VENOUS PRESSURES

Under steady-state conditions, as the pressure-flow equation indicates, it is the pressure in the dural sinuses that is important so far as the CSF pressure is concerned; the pressure in these vessels which are supported within the fibrous structure of the dura is virtually immune from the effects of changes in intracranial volume, and thus they are ideal vessels into which the fluid may be drained. This independence of the sinus pressure from the CSF pressure, which is such a valuable feature of the drainage mechanism, is easily shown by increasing the CSF pressure artificially through a cannula inserted into the subarachnoid space; thus Weed and Flexner (1933) showed that a rise to 450 mm H_2O or a fall to -300 mm H_2O were without effect on the torcular pressure in dogs. If the changes are induced rapidly, however, the diminished blood flow through the cranial blood vessels actually reduces the torcular pressure temporarily (Bedford, 1942). Thus, an acute rise in pressure of the veins in the subarachnoid space or nervous tissue will raise the CSF pressure, but this rise will not be transmitted to the dural sinuses, so that the effective pressure driving fluid out of the subarachnoid spaces will be increased; drainage will thus increase and the CSF pressure will fall. In so far as a raised venous pressure caused a rise in sinus pressure, of course the final situation would be a permanently raised CSF pressure. Raising the sinus pressure directly, e.g. by occlusion of the jugulars, will have both acute and chronic effects on the CSF pressure; thus the tendency to drainage will be reduced, causing an increase in pressure; furthermore, the raised sinus pressure may well increase the cranial venous pressure and this will be transmitted to the cerebrospinal fluid. The ultimate effect of jugular occlusion will depend on the level at which the sinus pressure settles; if adequate compensation is possible, i.e. if it returns to its pre-compression value, then the CSF pressure will return to normal. In fact, Bedford (1935) found that, in the dog, the CSF pressure did return to almost its normal value, and this was associated with a return of the torcular venous pressure to its original level.

GENERAL CONCLUSION

This review of the cerebrospinal fluid, regarded from the limited aspect of the maintenance of the cerebrospinal fluid pressure, has brought into prominence the dominant role of the resistance in the drainage channels in determining a pathologically raised pressure; our knowledge of the physiology of secretion of the fluid has pointed to the possibility of reduction in the rate as a possible means of relieving a pathologically raised pressure, and the study of the steady-state pressure-flow equation indicates that this reduced rate of secretion will be accompanied by a fall in CSF pressure, the higher this pressure the more dramatic will be the effect, since flow will depend on $P_{CSF} - P_{Sinus}$; the greater the value of P_{CSF} the nearer will this difference be equal to P_{CSF}. Vascular factors will have a permanent effect on CSF pressure only to the extent that they are able to influence drainage rate and secretion rate; thus we have seen how an acute rise in venous pressure may produce a rise in CSF pressure by simple transmission of pressure through the thin and easily distensible walls of the veins; this raised CSF pressure will be compensated, however, by the increased drainage that must occur, so that unless the sinus pressure increases, the CSF pressure returns to its original value. Secretion rate may be reduced by the ischaemia of hypothermia or of local noradrenaline, but whether there are normal fluctuations in secretion rate due to normal fluctuations in blood flow through the choroid plexuses cannot be stated with certainty.

REFERENCES

BEDFORD T.H.B. (1935) *Brain* **58,** 427
BEDFORD T.H.B. (1942) *J. Physiol. (Lond.)* **101,** 362
BERING E.A. (1962) *J. Neurosurg.* **19,** 405
BERING E.A. and SATO O. (1963) *J. Neurosurg.* **20,** 1050
BRADBURY M.W.B. and DAVSON H. (1964) *J. Physiol. (Lond.)* **170,** 195
BRADBURY M.W.B. and LATHAM W. (1965) *J. Physiol. (Lond.)* **181,** 785
DANDY W.E. (1919) *Ann. Surg.* **70,** 129
DANDY W.E. and BLACKFAN K.D. (1914) *Am. J. Dis. Child.* **8,** 406
DAVSON H. (1956) In *Physiology of the Ocular and Cerebrospinal Fluids.* Churchill, London
DAVSON H. (1958) In *The Cerebrospinal Fluid,* p. 189. Ciba Foundation Symposium. Churchill, London
DAVSON H. and BRADBURY M. (1964) *Symp. Soc. exp. Biol.* **19,** 349
DAVSON H. and LUCK C.P. (1957) *J. Physiol. (Lond.)* **137,** 279
DAVSON H. and POLLAY M. (1963) *J. Physiol. (Lond.)* **167,** 239
DAVSON H. and SPAZIANI E. (1962) *Expl. Neurol.* **6,** 118
ELMAN R. (1923) *Johns Hopkins Hosp. Bull.* **34,** 99
FISHMAN R.A. (1959) *J. clin. Invest.* **38,** 1698
FLEXNER L.B. (1933) *Am. J. Physiol.* **106,** 201
FLEXNER L.B. and WINTERS H. (1932) *Am. J. Physiol.* **101,** 697
FRAZIER C.H. and PEET M.M. (1914) *Am. J. Physiol.* **35,** 268
GRUNDY H.F. (1962) *J. Physiol. (Lond.)* **163,** 457
HEISEY S.R., HELD D. and PAPPENHEIMER J.R. (1962) *Am. J. Physiol.* **203,** 775
JAVID M., GILBOE D. and CESARIO T. (1964) *J. Neurosurg.* **21,** 1059
KOCH A. and WOODBURY D.M. (1960) *Am. J. Physiol.* **198,** 434
MELLANBY E. (1941) *J. Physiol. (Lond.)* **99,** 467
MOORE L.A. and SYKES J.F. (1941) *Am. J. Physiol.* **134,** 436
OPPELT W.W., PATLAK C.S. and RALL D.P. (1964) *Am. J. Physiol.* **206,** 247
PAPPENHEIMER J.R., FENCL V., HEISEY S.R. and HELD D. (1965) *Am. J. Physiol.* **208,** 436

PAPPENHEIMER J.R., HEISEY S.R., JORDAN E.F. and DOWNER J.DE C. (1962) *Am. J. Physiol.* **203,** 763

PAPPIUS H.M. and DAYES L.A. (1965) *Archs Neurol. (Chicago)* **13,** 395

POLLAY M. and CURL F. (1967) *Am. J. Physiol.* **213,** 1031

POST R.L. and JOLLY P.C. (1957) *Biochim. Biophys. Acta* **25,** 118

REED D.J. and WOODBURY D.M. (1962) *J. Physiol. (Lond.)* **164,** 265

RIESELBACH R.E., DI CHIRO G., FREIREICH E.J. and RALL D.P. (1962) *New Engl. J. Med.* **267,** 1273

ROSOMOFF H.L. (1961) *J. Neurosurg.* **18,** 753

ROSOMOFF H.L. (1962) *J. Neurosurg.* **19,** 859

ROTHMAN A.R., FREIREICH E.J., GASKINS J.R., PATLAK C.S. and RALL D.P. (1961) *Am. J. Physiol.* **201,** 1145

SCHURR P.H., MCLAURIN R.L. and INGRAHAM F.D. (1953) *J. Neurosurg.* **10,** 515

SEGAL M. and DAVSON H. (1968) In preparation

SKOU J.C. (1957) *Biochim. Biophys. Acta* **23,** 394

SKOU J.C. (1960) *Biochim. Biophys. Acta* **42,** 6

SMYTHE L., SMYTHE G. and SETTLAGE P. (1950) *J. Neuropath. exp. Neurol.* **9,** 438

SWEET W.H., SELVERSTONE B., SOLLOWAY S. and STETTEN D. (1950) In *American College of Surgeons Forum*, p. 376. Saunders, Philadelphia

TSCHIRGI R.D., FROST R.W. and TAYLOR J.L. (1954) *Proc. Soc. exp. Biol. Med.* **87,** 373

TUBIANA M., BENDA P. and CONSTANS J. (1951) *Revue neurol.* **85,** 17

VAN WART C.A., DUPONT J.R. and KRAINTZ L. (1960) *Proc. Soc. exp. Biol. Med.* **103,** 708

VAN WART C.A., DUPONT J.R. and KRAINTZ L. (1961) *Proc. Soc. exp. Biol. Med.* **106,** 113

VATES T.S., BONTING S.L. and OPPELT W.W. (1964) *Am. J. Physiol.* **206,** 1165

WEED L.H. (1935) *Brain* **58,** 383

WEED L.H. and FLEXNER L.B. (1933) *Am. J. Physiol.* **105,** 266

WELCH K. (1963) *Am. J. Physiol.* **205,** 617

WELCH K. and FRIEDMAN V. (1960) *Brain* **83,** 454

WELCH K. and POLLAY M. (1963) *Anat. Rec.* **145,** 43

WOOLLAM D.H. and MILLEN J.W. (1958) In *The Cerebrospinal Fluid*, p. 124. Ciba Symposium. Churchill, London

CHAPTER SEVENTEEN

GENETICALLY DETERMINED NEUROLOGICAL DISEASES IN CHILDREN

JOHN WILSON

I think it is most fitting that there is at least one genetically oriented chapter in the present series, since a substantial proportion of the neurological disorders seen in our community at all ages are genetically determined more or less directly, and in the past two decades the metabolic basis of a few of them has been recognized. Not only has our understanding of the mechanism of individually rare diseases been thus enlarged, but more important, their study has thrown light on the aetiopathogenesis of other less rare neurological and non-neurological maladies.

As you know, neurologists have long displayed a keen interest in heredofamilial diseases and the literature of over a century is replete with numerous examples. It is worth reflecting that careful descriptions of a number of such disorders were presented well before Mendel's observations and hypotheses were publicized at the turn of the century. Dominantly inherited conditions constitute an unrepresentatively high proportion of cases described in the older literature simply because they were easier to recognize, being usually present in succeeding generations, but it is well to remember that there are many more recessively than dominantly inherited neurological conditions, and in a place like the National Hospital for Nervous Diseases there is almost certainly a substantial number of patients suffering from recessive disorders, many unrecognized as such. Careful inquiry about consanguinity and the health of siblings is extremely important, but obviously does not invariably disclose genetically relevant information.

In this chapter and Chapter 18 are reviewed the clinical, genetical and biochemical aspects of what we hope is a representative group of neurological disorders, especially those presenting in childhood. The conditions have been selected partly on the basis of their relative frequency, partly because of certain therapeutic implications, partly because of their topical interest. Whilst it is axiomatic that there is a sense in which all diseases are metabolic in character, at the cellular level if not in the conventional sense, it is not yet justifiable to claim that all heredo-degenerative neurological diseases are metabolically determined, although this might well be so. Nevertheless, the general proposition that recessive genes probably act through a specific defect in enzymic activity is certainly applicable to a considerable number of neurological disorders, and doubtless the next decade will see many more examples. I do not, however, subscribe to Becker's view (Becker, 1967) that whereas recessive diseases exert their effect through enzymic abnormalities, dominant genes are manifested as structural disorders.

CLASSIFICATION

For the purpose of this discussion, I have arbitrarily divided heredo-degenerative disorders into three broad groups:

1   Diseases of intracellular storage
2   Abiotrophies
3   Diseases secondary to a generalized metabolic disturbance

In the first group, a derangement of cell function results from the local accumulation of a metabolite, for example lipid in Tay-Sachs disease, mucopolysaccharide in Hurler's syndrome, or carbohydrate in glycogen-storage diseases.

Abiotrophy was Gowers's term to describe the phenomenon whereby, after a period of apparently normal development and differentiation lasting sometimes many years, the patient would be afflicted by an obviously progressive disorder. Although not originally intended, the term has come by usage to imply that parts of the neuraxis are inherently defective structurally and become prematurely effete. This hypothesis is remarkably devoid of a shred of supporting evidence except the natural history. Such diseases as Friedreich's disease, Huntington's chorea and Leber's disease are still included in this group, although others, for example the late-onset leucodystrophies which once were so classified, are now more satisfactorily included as structurally localized metabolic disorders. I would expect this group to diminish in size as our ideas on pathogenesis become clarified.

In those diseases secondary to a generalized metabolic disturbance, there is probably a good deal of overlap with the previously described groups. In diseases such as phenylketonuria, galactosaemia, hereditary fructose intolerance and porphyria, a generalized metabolic disturbance has a substantial impact on neuraxial function and often structure.

It must be emphasized that these subdivisions are highly arbitrary and their boundaries are defined by our ignorance rather than by our knowledge. But in providing a framework for this discussion, as a clinician I want to promulgate the view that with the elaboration of the techniques of analytical and enzymic biochemistry, our understanding of genetically determined neurological disorders will be vastly enhanced in the next two decades.

The diseases to be reviewed are as follows:

DISEASES OF CELLULAR STORAGE—THE SPINGOLIPIDOSES
Neurolipidoses
Tay-Sachs disease
Cerebro-macular degeneration
Niemann-Pick's disease
Gaucher's disease

Leucodystrophies
Metachromatic leucodystrophy
Krabbe's globoid cell leucodystrophy

ABIOTROPHIES
Friedreich's ataxia
Huntington's chorea

GENERALIZED METABOLIC DISORDERS
Lipid
Refsum's disease
a-β-lipoproteinaemia (Bassen-Kornzweig)

Purine
Hyperuricaemia and athetosis

?Metal ?protein
Hepatolenticular degeneration

Carbohydrate
Leigh's infantile necrotizing encephalomyelopathy
Hereditary fructose intolerance

Excluded from the list for reasons of brevity are such conditions as the myopathies, porphyria, pyridoxine-dependent fits, and kernicterus due either to rhesus incompatibility or glucose-6-phosphate dehydrogenase deficiency, though clearly the biochemically oriented geneticist has at least as much of importance to say about these disorders (if not more so) as the neuropathologist who pontificates over the last rites.

DISEASES OF CELLULAR STORAGE—
THE SPHINGOLIPIDOSES

The usual clinico-pathological differentiation of the sphingolipidoses into the neurolipidoses and the leucodystrophies depends on the relative degrees of involvement of grey and white matter.

In the former, primary involvement of the cerebral cortex is accompanied by accumulation of lipids intraneuronally, leading to disruption of cell function and structure. In the leucodystrophies, the white matter appears shrivelled and sclerotic at autopsy whilst the cortex may appear relatively normal on naked-eye examination. It is now recognized, however, that although the anatomical distribution of the principal lesions differ, the pathogenesis of the two groups of disorders is broadly similar.

TAY-SACHS DISEASE (Tay, 1881; Sachs, 1887)

This disease is inherited as an autosomal recessive condition, and usually presents in the first year of life, often before the age of 6 months.

Early diagnosis may be difficult in this as in other diseases of this group unless the condition is suspected for some reason, e.g. a previously affected sibling.

A featureless failure to thrive may give place to definite evidence of visual failure. There is an insidious arrest of development and then regression, with the later onset of major fits. Outbursts of uncontrolled and unnatural laughter may occur, so-called gelastic seizures: both they and more typical seizures respond poorly to anti-convulsants. The composite picture is progressive—and the child dies of inanition or respiratory tract infection, sometimes in as little as 3–6 months from the onset, usually within 2 years.

A cardinal clinical feature sooner or later is the so-called cherry-red spot at the macula. This is not specific for Tay-Sachs disease but is also found in a substantial minority of patients with Niemann-Pick's and Gaucher's disease. The appearance is due to the contrast between the pink spot at the macula where the vascular choroid appears through the thinned retina, and the grey appearance around this resulting from the heavy deposition of lipid in the ganglion cells, and associated degeneration.

An excessive startle response to auditory stimuli is also quite character-istic. The infant is usually hypotonic but has exaggerated tendon jerks.

Clinically there is no evidence of dysfunction outside the central nervous system, but it is probable that in patients surviving long enough there is abnormal lipid accumulation demonstrable elsewhere in the reti-culoendothelial system (Crome, personal communication).

PATHOLOGY (Aronson, Volk and Epstein, 1955; Aronson, Lewitan, Rabiner, Epstein and Volk, 1958; Aronson and Volk, 1962; Terry and Weiss, 1963)

At autopsy, the brain may be atrophied, particularly in the cerebellum, but in some cases lipid accumulation is sufficiently excessive to cause the brain to be increased in size. Microscopically, ballooning of neurones due to lipid accumulation also affects astrocytes and microglia, partly through phagocytosis of degenerating axons. Meningeal thickening with lepto-meningeal foam-cell deposition is seen in advanced cases.

Intraneuronal deposition of lipid is also seen in the ganglion cells of intestinal mucosa, and therein lies the diagnostic value of rectal biopsy in this and allied conditions (Bodian and Lake, 1963) where histological abnormality may presage clinical involvement (Brett and Berry, 1967).

Ultrastructural studies

Although mitochondria, nuclei, ribosomes and endoplasmic reticulum appear normal in ganglion cells, the cytoplasm in these cells, as well as in glia, astrocytes and axis cylinders contain numerous round or oval membranous cytoplasmic bodies (m.c.b.). These have a lamellar appear-ance and are composed of a variety of lipids, mainly ganglioside (Korey *et al*, 1963a and b; Samuels, Korey, Gonatas, Terry and Weiss, 1963; Terry and Korey, 1963; Terry and Weiss, 1963).

OTHER GANGLIOSIDOSES

Ranged alongside this well-defined disease of ganglioside storage are several related but distinct disorders also due to ganglioside accumulation in the central nervous system. These conditions were misleadingly known as the *amaurotic family idiocies*, but this lamentable term is gradually being abandoned.

These diseases have in common with Tay-Sachs disease progressive psychomotor dysfunction and seizures, but they are distinct genetically, clinically and chemically from that condition. Their differentiation into separate syndromes has been based on age of onset rather than any more fundamental criteria, and they are usually known eponymously as Bielschowsky (1914), the late infantile; Spielmeyer (1908), Vogt (1905) or Batten (1903), the juvenile form (the name chosen depending upon one's patriotic loyalties); and the Kufs (1925) or adult forms.

Their individual, as opposed to their group identity is less certain, and some workers believe that they are pathogenetically the same disorder, differing only in age of onset, in a way not unknown in the spectral variation of neurological disease.

At the Hospital for Sick Children we see many more examples of the disease conforming to descriptions of the late infantile type than to any other, and it is noteworthy that the majority of our patients are non-Jewish.

Although psychomotor development is usually normal during infancy, a minority of patients have unquestionable delay in the first year, although actual regression is not usually recognized until the second year.

In this group of children, inco-ordination and ataxia may result from the co-existent cerebellar dysfunction and myoclonus, usually with minor seizures. Major seizures may supervene, and not until considerably later does visual deterioration often of cortical origin occur. In some patients, however, it is possible to see minor pigmentary changes in the region of the macula.

Pyramidal signs in the limbs may tempt the incautious to make a diagnosis of cerebral birth injury, but a careful history should leave one in no doubt about the progressive nature of the disorder.

EEG changes parallel the evolving character of the disease, and these have been carefully studied by Dr Pampiglione at the Hospital for Sick Children. He has described a most interesting response to photic stimulation in children with the late-infantile type of ganglioside neurolipidosis. This comprises an evoked response produced at relatively low flicker rates and recurring at an iodsyncratic frequency at higher flicker rates (Pampiglione, 1961). This response is sufficiently characteristic to be of diagnostic value, but is not seen in Tay-Sachs disease, perhaps because blindness is such an early feature. Nor is this EEG appearance seen in other degenerative brain disorders.

Unfortunately, none of the late-onset examples of ganglioside neurolipidosis has come Dr Pampiglione's way for investigation and therefore its intra-group specificity in these disorders is unproven.

Radiological investigations are not particularly helpful: they may show cerebral and cerebellar atrophy.

Genetics

All the diseases of this group are inherited as autosomal recessive conditions. Tay-Sachs disease is genetically distinct from the later-onset cases in which, as mentioned earlier, the inheritance has not been studied sufficiently to allow firm genetic distinctions to be made. Although Tay-Sachs disease occurs more commonly in Jewish infants than in non-Jewish infants (12 and 0.2/100,000 births respectively (Kozinn, Wiener and Cohen, 1957)) the disease appears to be identical in both groups.

NIEMANN-PICK'S DISEASE (Niemann, 1914; Pick, 1927) AND
GAUCHER'S DISEASE (Gaucher, 1882)

Other diseases in which there is histological and chemical evidence of lipid accumulation in neurones result in a similar clinical picture—regression, fits, increasing pyramidal signs, but in addition there is visceral involvement of the reticuloendothelial system, liver and spleen. In Niemann-Pick's disease there is a gross accumulation of sphingomyelin in liver and spleen, whilst in Gaucher's disease cerebroside deposition is particularly evident in the spleen. Similar widespread, though not so conspicuous, visceral involvement is not unknown in the disorders already discussed.

Although I think a holistic view of pathogenesis is fully justified for several reasons, it must be emphasized that in these diseases a superficial similarity is compatible with a variety of different processes. A large number of distinct (though related) processes may occur and produce qualitatively similar end-results, differing clinically in the relative degrees of visceral and neurological involvement, and differing fundamentally in their chemical and presumably enzymic origin. There are numerous examples published of so-called 'intermediate' forms which seem to provide bridges between the main groups of disorders, and whilst considering them as transitional forms may prove a useful *aide-mémoire*, they do represent independent diseases which need to be characterized chemically and enzymically in their own right.

LEUCODYSTROPHIES

As mentioned earlier, although these conditions were considered to be degenerative disorders confined to the white matter, it is now recognized that some of them result, like the neurolipidoses, from primary abnormalities of lipid metabolism, and grey-matter involvement does indeed occur.

The best known are Greenfield's metachromatic leucodystrophy Greenfield, 1933) and globoid cell leucodystrophy (Krabbe, 1916) which are now clearly distinguished from sudanophilic diffuse sclerosis and Schilder's disease with which they were formerly grouped.

METACHROMATIC LEUCODYSTROPHY

As with the ganglioside neurolipidoses, age-specific syndromes are described. Although the late-infantile form (van Bogaert and Scholz, 1932; Greenfield, 1933) is the commonest and apparently a genetically homogeneous autosomal recessive condition, the later-onset forms are less well characterized independently, perhaps because of their rarity.

The late-infantile disease may present as a cerebellar ataxia (Thieffry, 1962), but sooner or later intellectual deterioration supervenes.

Fits occur in a substantial minority of patients, but a characteristic though not invariable feature is peripheral nerve involvement evidenced by weakness, depression of tendon jerks and slowing of nerve conduction (Fullerton, 1964). Brisk toe jerks mediated as a cutaneous reflex may contrast with the distal loss of tendon reflexes, and this association has been claimed to be pathognomonic.

In the adult patients, a psychotic illness with depressive and schizoid features may presage more obvious dementia, but the tempo of the illness is much slower than in children.

Neuropathology

The white matter is severely involved in the brain and is often shrunken, hard and even cavitated.

On microscopic examination there is gross demyelination which may involve peripheral nerves in a segmental fashion as well as cerebral white matter. Spherical masses of metachromatically staining lipid accumulate in macrophages and also lie free in the brain substance. They may also accumulate in oligodendrocytes in areas where myelin is apparently normal. In peripheral nerves metachromatically staining lipid may be seen in Schwann cells.

This material may also be deposited in the epithelial cytoplasm of the renal convoluted tubules (Jatzkewitz, 1958; Austin, 1959) and collecting tubules, and can be demonstrated intracellularly in the freshly spun urinary deposit (Austin, 1957a and b; Lake, 1965). There is no evidence, however, of impaired renal function. Similarly, accumulation occurs in the liver, gall-bladder (Hagberg and Svennerholm, 1960) and various endocrine tissues (Witte, 1921; Denny-Brown, Richardson and Cohen, 1962; Hagberg, Sourander and Svennerholm, 1962).

Urinary examination is extremely valuable in diagnosis, but there must be great care to prepare the spun deposit promptly after collection. Only *intracellularly* staining material is significant. For a description of a method which gives consistently successful results I recommend Lake's original paper (Lake, 1965).

GLOBOID CELL LEUCODYSTROPHY (Krabbe, 1916)
This disease is inherited as an autosomal recessive condition, and presents both distinctive clinical and pathological changes.

The onset is in infancy in most patients, sometimes presenting as early as 3 months, with vomiting and failure to thrive. Irritability often becomes a prominent feature, the infant crying inconsolably whether disturbed or not. Although blindness with optic atrophy may be discovered within a few weeks, deafness is not a feature, at least in the early stages. Nor is the hyperacusis of Tay-Sachs disease reproduced in this condition. At first the clinical picture is dominated by increasing tonus of the limbs with exaggerated tendon jerks, so that a diagnosis of quadriplegic cerebral palsy might be made. The loss of skills quickly demonstrates that this diagnosis is untenable. Later the hypertonus seems to be striatal in character and dystonic postures and movements are seen. Neck retraction and opisthotonic seizures are not uncommon.

Recently we have realized that, in some patients, as the disease progresses tendon reflexes are lost, and there is marked slowing of nerve conduction (Dunn, Lake and Wilson, 1968).

The malady is uniformly fatal, and survival is rarely as much as 2 years.

Pathology
The brain may be either shrunken or distended, with severe white-matter involvement. Histological examination of the brain reveals not only widespread loss of myelin, but also distended cells, probably astrocytes with a foam-like appearance, so-called globoid cells or gemistocytes.

Corresponding with the clinical and electrophysiological evidence of peripheral neuropathy, segmental demyelination has recently been described by Lake (1968). In addition, the increased alkaline phosphatase activity histochemically seen in ganglion cells and neurones also occurs in peripheral nerves. Thus, in this disease, as in metachromatic leucodystrophy, myelin destruction in peripheral nerve coexists with the central white-matter changes.

OTHER SUDANOPHILIC SCLEROSES
It is clear that there is a miscellany of other leucodystrophic disorders which present a non-specific clinical picture but whose neuropathological

features are sufficiently distinctive to merit description from time to time in the literature. Unfortunately, such conditions are not sufficiently common to allow of systematic genetic or biochemical study yet, although doubtless one might reasonably expect some of these disorders to represent genetically determined primary metabolic disturbances.

ABIOTROPHIES

FRIEDREICH'S DISEASE (Friedreich, 1863a, b, c, d)

The clinical evidence of Friedreich's disease usually develops insidiously in childhood. An initially ungainly walk becomes frankly ataxic, and as the disease progresses inco-ordination of cerebellar type is superimposed on sensory ataxia.

High-arched feet, thoracic scoliosis, loss of tendon jerks and impaired proprioception are the other features which comprise this distinctive condition. In advanced cases, dementia and optic atrophy are commonly seen. There is also an increased incidence of diabetes mellitus, although this is uncommon in childhood cases.

Although it is the current trend to regard abiotrophic disorders as the clinical expression of biochemical errors, the direct evidence to support this view in Friedreich's disease is negligible. A satisfactory biochemical hypothesis must take account not only of the varying age of onset and time course of the disease, but also explain the varying clinical manifestations in different individuals.

The majority of cases appear to conform to the criteria of autosomal recessive inheritance, but the existence of 'formes frustes', which are not uncommon, make too literal interpretation of pedigree data unwise. Moreover, certain studies (Roth, 1948) suggest that, for example, peroneal muscular atrophy and Friedreich's ataxia may be clinical variants of the same underlying genetic abnormality. These problems will remain unsolved until discriminatory parameters, perhaps biochemical, which consistently identify the genetic abnormality, are available.

HUNTINGTON'S CHOREA

This well-known condition usually afflicts adults in their fourth and fifth decade, but, of course, insidious personality changes may be considered retrospectively to have been present for some time. Not only is there a good deal of variation in the clinical picture, but also in the age of onset, and the occurrence of the disease in childhood is well recognized.

Disinhibited, fatuous behaviour or even a psychosis may presage the more florid features of the disease, with grimacing and choreiform or

S

dystonic movements. This is the most commonly recognized pattern of the disorder in adults. In a minority of adults there is a so-called rigid form of the disease (Westphal variant) in which akinesia, or parkinsonian-type rigidity with action and intention tremor, may be present long before dementia is prominent.

It is clear that the two forms are not distinct genetically, examples of both occurring in the same pedigree following the usual autosomal dominant pattern of inheritance. The absence of antecedent family history (which is not due to ignorance, suppression of information, or premature death) almost certainly represents the occurrence of mutations, and should not prevent one considering the diagnosis.

Under 10 per cent of patients with Huntington's chorea are children (Bruyn, 1967). Although the typical picture of slowly progressive dementia and chorea is seen in childhood, it is unusual, and the disease is more likely to appear and to progress dramatically. A hemiparesis of subacute onset may be the first hint of neurological disease, whilst the obtrusion of cerebellar symptoms and signs may erroneously suggest a diagnosis of cerebellar degeneration.

This gives place more or less rapidly to rigidity and a parkinsonian-type picture. Dementia occurs early. 'Pseudo-ophthalmoplegia' similar in certain respects to the oculomotor apraxia of Cogan (1953) occurs as part of more widespread difficulty of voluntary facial movements, which may be properly called facial apraxia.

Major fits and myoclonus are common in children suffering from this disease, and respond poorly to medication. In the pre-terminal stages ictal crises of rigidity with profuse sweating and precipitately falling body temperatures to sub-normal levels may produce critically dehydrating fluid loss.

It seems certain that the juvenile form is genetically homogeneous, with the adult cases as an autosomal dominant, but one curious and un-explained difference is that juvenile patients inherit this disease predominantly from their fathers, affected males occurring more than three times as often as affected females among the parents of such children (Merritt, Conneally, Rahmann and Drew, 1967). Unfortunately, there is not sufficient data about fertility rates, ages of marriage, life expectancy and socio-economic factors among parents of affected children to decide why this should be so. The difference could be explained by a higher mortality rate in children of affected mothers (who tend to be promiscuous anyway) than in children whose fathers are affected, and who are more likely to survive to develop the disease.

GENERALIZED METABOLIC DISORDERS

REFSUM'S DISEASE

Refsum's disease undoubtedly occurs in its fully developed form in child-hood. Night blindness, atypical retinitis pigmentosa, nerve deafness, ataxia, peripheral neuropathy and cardiac arrhythmia have all been described. Refsum and his colleagues (Refsum, Salomonsen and Skatvedt, 1949) described the condition in four children of whom three were affected by the age of 7 years, and one at the age of 4. The little girl described by Heycock and Wilson (1958) was also affected at the age of 4 years, and later developed diabetes mellitus. The condition is inherited in adults as an autosomal recessive and seems genetically homogenous with the child-hood cases. In both children and adults the disease may pursue a fluctuating course: hearing loss may occur suddenly or gradually, whilst tendon jerks, which are usually lost early, may transiently reappear. Although peripheral nerves may not be palpably thickened, there is histological evidence of a hypertrophic neuritis. Longevity is usually considerably curtailed and death usually results from acute cardiac arrhythmia.

A-β-LIPOPROTEINAEMIA (Bassen and Kornzweig, 1950)

This is a condition which has a slight clinical resemblance to Refsum's syndrome, and may be confused with it.

There are mild pigmentary changes around the macula, but cerebellar ataxia is unaccompanied by peripheral nerve involvement. A malabsorption syndrome in the very young mimics coeliac disease, but is not gluten-sensitive. In the peripheral blood, numbers of red blood cells have a prickly, mulberry-like appearance, so-called acanthocytes, but these are not specific for the condition.

HYPERURICAEMIA, ATHETOSIS, MENTAL RETARDATION

Surely one of the most bizarre neurological syndromes which has been described in recent years is that associated with a raised blood uric acid. The syndrome was first recognized by Catel and Schmidt (1959) and later by Riley (1960).

The distinctive features of the syndrome are mental retardation, choreoathetosis and self-mutilation, occurring in early childhood.

Certain patients have been somewhat hypotonic at birth, but without any hint of trouble until later during the first year when motor development has been delayed. By the end of the first year, and afterwards, increasing evidence of choreoathetosis, dystonic postures with varying hypertonus and some jerky movements, has led usually to a diagnosis of athetoid cerebral palsy. But by the time the deciduous dentition is fully established during the second and third years, the most distressing aspect of the disease, self-mutilation, occurs. Biting of the hands and forearms is not uncommonly seen in retarded children; it then usually involves the dorsum of the hand or metacarpo-phalangeal joints or forearms, with occasional biting of the lips, but not of the degree seen in the condition now under discussion. In this disease, the lower lip may be completely bitten away, and the terminal phalanges may be gnawed or lost through secondary infection and gangrene.

Besides this compulsive behaviour, some of the children bang themselves sufficiently violently to break their noses, and some American illustrations show patients without lips or nasal septa.

Aggressive outbursts towards others may include obscene abuse as well as physical violence.

Some of these children are apparently much happier when arms and body are restrained with straps and splints, and hands embedded in generous layers of padding. Their removal results in the resumption of finger-biting, almost immediately.

An element of progression in the neurological condition seems to be lacking in the patients so far described, although longevity is curtailed either by death from infection or renal disease. Self-mutilation is not, however, invariable.

The elevated blood uric acid which is a constant feature of the disease is not usually associated with the more typical evidence of gout such as tophi and arthritis, although this is how Riley's (1960) patient presented. In the patient described by Sass, Habashi and Dexter (1965) tophi and arthritis occurred in the presence of markedly impaired renal function, but there was no gross renal involvement in Riley's case.

Except for one of the cases of Hoefnagel, Andrew, Mireault and Berndt (1965) there is no family history of gout. Even in Hoefnagel's family a diagnosis of gout in one father must remain very much in doubt when the serum uric acid level during a single acute attack of arthritis was only 6.2 mg per 100 ml.

The mechanism of the disease is in doubt. Although, as you will hear later, there is a well-defined abnormality of purine metabolism, clinically and genetically it does not resemble gout in its usual form. Paraplegia, mononeuritis and 'gouty headaches' have been described in gout sufferers, and one might question whether or not the hyperkinetic restlessness of the acute attack of gouty arthritis reflects cerebral dysfunction secondary to hyperuricaemia. But I do not really think that there is any question of clinical heterogeneity in a genetically homogeneous condition here. Adult gouty hyperuricaemia is probably a dominantly or perhaps polygenically inherited condition, whilst nearly all the evidence so far accumulated points to the athetoid syndrome being inherited as a sex-linked recessive disorder.

Pathology
Neuropathological studies of this disease have so far been very few. The haemorrhagic, vascular and degenerative changes in the brain of the uraemic patient of Sass *et al* (1965) are the well-recognized changes accompanying renal failure. In a recently described patient who died after a very brief febrile illness (Partington and Hennen, 1967) there were surprisingly few changes seen in the brain. In frozen sections occasional minute birefringent crystals could be seen in a perivascular distribution, but efforts to identify them as uric acid failed.

Treatment
Treatment with probenecid to increase uric-acid excretion, and allopurinol to reduce uric-acid production, have been successful in so far as they reduced the blood concentration of uric acid, but there is some doubt about whether there was any clinical benefit or not.

I believe that this disease is distinct from that customarily known as gout, and I think that time-honoured term should be reserved for its evocative, if apocryphal, association with high living.

HEPATOLENTICULAR DEGENERATION
There has been a sustained interest in this disease at this hospital ever since Dr Kinnier Wilson, formerly a physician here, described the condition in 1912 in a patient whom he had seen (Wilson, 1912). The association of hepatic cirrhosis and a choreiform syndrome was described, and has subsequently been amply confirmed. Latterly the metabolic studies of Professor Cumings have sustained and enlarged the interest in the condition.

Wilson's patient was a young man whose main clinical features were muscular rigidity, tremor and grimacing. Hepatic cirrhosis and lenticular degeneration were found at autopsy, but this particular presentation of the disease is less common than the occurrence of dysarthria, dysphagia and rigidity. At least half of the cases now diagnosed present in late childhood or early adolescence. Bulbar symptoms may be very advanced before any others develop: but clearly in an evolving disease the clinical picture varies with its duration. Tendon jerks are usually normal.

A nearly pure parkinsonian syndrome is sometimes seen, or a progressive choreoathetosis.

In advanced disease, the patient is unable to speak, and there is loss of emotional control.

A myotonoid reaction has been described but this may represent a dyspraxic difficulty: a truly myopathic explanation seems very improbable.

In late cases, dystonic spasms with preservation of consciousness have been recognized for some time and have been compared with tetanus. In fact Sir William Gowers presented a paper 'On tetanoid chorea and its association with cirrhosis of the liver' from this hospital in 1906. His patients certainly resembled Wilson's.

In the syndrome so far described, clinical evidence of hepatic involvement is inconspicuous; in early childhood, however, this malady usually presents as jaundice which may have both hepatic and haemolytic components. Clinically the distinction from hepatitis is sometimes difficult but it is obviously important to recognize the possibility. The liver, if enlarged, is usually non-tender.

The Kayser-Fleischer ring of corneal pigmentation may be recognizable at an early stage of the disease, and its presence at any stage is pathognomonic.

Genetics

The disease is inherited as an autosomal recessive, but it is not known whether the wide range of age of onset is due to genetic heterogeneity or to the presence of modifying genes.

INFANTILE NECROTIZING ENCEPHALOMYELOPATHY (Leigh, 1951)
This disease, although rare, is worth discussing not only for its intrinsic interest, but also because of the therapeutic implications of some recent studies.

Although the 8-month-old infant whose autopsy was described by Leigh (1951) died after a rapidly progressive but rather featureless illness with arreflexic hypertonic limbs, optic atrophy, deafness and pupillary abnormalities, the neuropathological changes were quite striking. There was diffuse vascular injection of the leptomeninges, and the white matter generally appeared grey. Grey-brown naked-eye discoloration of thalamus, midbrain and posterior columns of the cervical and thoracic cord

were seen on microscopical examination to be the site of profuse pro-
liferative changes in small vessels. A variety of degenerative changes were
seen in the neurones in these areas, with intense glial proliferation in areas
of softening.

Although Leigh compared the changes he saw with those of Wernicke's
encephalopathy, he and others have pointed out that unlike that condition,
the mammillary bodies were not affected, whilst the spinal cord was.

This is a rare disease, but a number of infants and children apparently
suffering from the same condition neuropathologically have since been
reported. The clinical picture can vary—ataxia, external and internal
ophthalmoplegia, and pyramidal signs have all been described, together
with a fluctuating course, in which unexplained episodes of vomiting,
diarrhoea, sweating and hyperventilation may punctuate and often
terminate the neurological illness.

The age of onset varies in infancy and early childhood, but there is not
as yet sufficient evidence to support the suggestion of Lakke, Ebels and
ten Thye (1967) that there are infantile and juvenile forms of the disease.

Biochemical studies have shown that Leigh's comparison with Wernicke's encephalopathy was not fanciful, and the recent promising results
of treatment with lipoic acid further support this view (Clayton, Dobbs
and Patrick, 1967). Similar success since has been claimed by other
workers (Clayton, personal communication) and naturally at the Hospital
for Sick Children we are trying to identify patients before the pathologist
does! Clearly, however, effective treatment in this and many other
degenerative diseases must be undertaken at an early stage before
irreparable neuraxial damage has occurred.

HEREDITARY FRUCTOSE INTOLERANCE

Chambers and Pratt (1956) first described this recessively inherited
metabolic disorder in which the ingestion of fructose or fructose-contain-
ing sugars, viz. sucrose, may cause severe hypoglycaemia acutely, or,
more insidiously, fatty infiltration and cirrhosis of the liver. Jaundice,
hepatic enlargement, failure to thrive and hypoglycaemic fits may occur
in this condition which usually presents in infancy at the time of the
introduction of mixed feeds.

Intolerance may apparently persist, though usually less intensely,
throughout life.

The intolerance is similar to that seen in galactosaemia, but in both
diseases, although severe hypoglycaemia may occur, the biochemical
abnormalities are quite distinct. Both conditions are apparently recessive.

Treatment is simple and highly effective—the withdrawal of sucrose and sweet things from the diet.

CONCLUSION

In terms of the sum total of human suffering, these diseases are rare, and their intensive study must seem to some a marvellously irrelevant and expensively subsidized hobby. This is a short view: in the long term, a fuller understanding of the mechanism of inherited disorders will be valuable in a much wider context. With the advanced technology of the biochemist and biophysicist now available, it seems opportune to concentrate more, not less, effort into the study of inherited neurological disease because with present knowledge the economics of prevention of these conditions through prophylaxis (as in Wilson's disease) and more precise genetic counselling is such that there is considerable merit in these efforts already.

REFERENCES

ARONSON S.M., LEWITAN A., RABINER A.M., EPSTEIN N. and VOLK B.W. (1958) *A.M.A. Archs Neurol. Psychiatry* **79,** 151

ARONSON S.M., VOLK B.W. and EPSTEIN N. (1955) *Am. J. Path.* **31,** 609

ARONSON S.M. and VOLK B.W. (1962) In *Cerebral Sphingolipidoses: A Symposium on Tay-Sachs' Disease and Allied Disorders*, p. 15. Ed. ARONSON S.M. and VOLK B.W. Academic Press, New York.

AUSTIN J.H. (1957a) *Neurology (Minneap.)* **7,** 415

AUSTIN J.H. (1957b) *Neurology (Minneap.)* **7,** 716

AUSTIN J.H. (1959) *Proc. Soc. exp. Biol. Med.* **100,** 361

BASSEN F.A. and KORNZWEIG A.L. (1950) *Blood* **5,** 381

BATTEN F.E. (1903) *Trans. ophthal. Soc. U.K.* **23,** 386

BECKER P.E. (1967) *Excerpta Med. International Congress Series* **154,** 18 (Second International Congress of Neuro-genetics and Neuro-ophtholmology, Montreal, 1967)

BIELSCHOWSKY M. (1914) *Dt. Z. NervHeilk.* **50,** 7

BODIAN M. and LAKE B.D. (1963) *Br. J. Surg.* **50,** 702

BRETT E.M. and BERRY C.L. (1967) *Brit. med. J.* **3,** 400

BRUYN G.W. (1967) *Excerpta Med. International Congress Series* **154,** 46. (Second International Congress of Neuro-genetics and Neuro-ophthalmology, Montreal, 1967)

CATEL W. and SCHMIDT J. (1959) *Dt. med. Wschr.* **84,** 2145

CHAMBERS R.A. and PRATT R.T.C. (1956) *Lancet* **ii,** 340

CLAYTON B.E., DOBBS R.H. and PATRICK A.D. (1967) *Archs Dis. Childh.* **42,** 467

COGAN D.C. (1953) *Am. J. Ophth.* **36,** 433

DENNY-BROWN D.E., RICHARDSON E.P. Jr and COHEN R.B. (1962) *New Engl. J. Med.* **267,** 1198

DUNN H., LAKE B.D. and WILSON J. (1968) In preparation

FRIEDREICH N. (1863a, b, c) *Virchows Arch. path. Anat. Physiol.* **26,** 391, 419, 433

FRIEDREICH N. (1863d) *Virchows Arch. path. Anat. Physiol.* **27,** 1

FULLERTON P.F. (1964) *J. Neurol. Neurosurg. Psychiat.* **27,** 100

GAUCHER P. (1882) Quoted in *The Metabolic Basis of Inherited Disease* (1966) 2nd ed., p. 565. Ed. FREDRICKSON D.S., STANBURY J.B. and WYNGAARDEN J.B. McGraw-Hill, New York

GOWERS W.R. (1906) *Rev. Neurol. Psychiat.* **9,** 249

GREENFIELD J.G. (1933) *J. Neurol. Psychopath.* **13,** 289

HAGBERG B., SOURANDER P. and SVENNERHOLM L. (1962) *Am. J. Dis. Child.* **104,** 644

HAGBERG B. and SVENNERHOLM L. (1960) *Acta paediat., Uppsala* **49,** 690

HEYCOCK J.B. and WILSON J. (1958) *Archs Dis. Childh.* **33,** 320

HOEFNAGEL D., ANDREW E.D., MIREAULT N.G. and BERNDT W.O. (1965) *New Engl. J. Med.* **273,** 130

JATZKEWITZ H. (1958) *Hoppe-Seyl. Z.* **311,** 279

KOREY S.R., GOMEZ C.J., STEIN A., GONATAS J., SUZUKI K., TERRY R.D. and WEISS M. (1963a) *J. Neuropath. exp. Neurol.* **22,** 2

KOREY S.R., GOMEZ C.J., STEIN A., GONATAS J., SUZUKI K., TERRY R.D. and WEISS M. (1963b) *J. Neuropath. exp. Neurol.* **22,** 10

KOZINN P.J., WIENER H. and COHEN P. (1957) *J. Pediat.* **51,** 58

KRABBE K. (1916) *Brain* **39,** 74

KUFS H. (1925) *Zentbl. ges. Neurol. Psychiat.* **95,** 169

LAKE B.D. (1965) *Archs Dis. Childh.* **40,** 284

LAKE B.D. (1968) *Nature (Lond.)* **217,** 171

LAKKE J.P.W.F., EBELS E.J. and TEN THYE O.J. (1967) *Archs Neurol., Chicago* **16,** 227

LEIGH D. (1951) *J. Neurol. Neurosurg. Psychiat.* **14,** 216

MERRITT A.D., CONNEALLY P.M., RAHMANN N.F. and DREW A.L. (1967) *Excerpta Med. International Congress Series* **154,** 45. (Second International Congress of Neuro-genetics and Neuro-ophthalmology, Montreal, 1967)

NIEMANN A. (1914) *Jb. Kinder Heilk* **79,** 1

PAMPIGLIONE G. (1961) *Soc. Graf. Romana* **1,** 53. (VII International Congress of Neurology, Rome, September 1961)

PARTINGTON M.W. and HENNEN B.K.E. (1967) *Develop. Med. Child. Neurol.* **9,** 563

PICK L. (1927) *Med. Klin.* **23,** 1483

REFSUM S., SALOMONSEN L. and SKATVEDT M. (1949) *J. Pediat.* **35,** 335

RILEY I.D. (1960) *Archs Dis. Childh.* **35,** 293

ROTH M. (1948) *Brain* **71,** 416

SACHS B. (1887) *J. nerv. ment. Dis.* **14,** 541

SAMUELS S., KOREY S.R., GONATAS J., TERRY R.D. and WEISS M. (1963) *J. Neuropath. exp. Neurol.* **22,** 81

SASS J.K., HABASHI H.H. and DEXTER R.A. (1965) *Archs Neurol., Chicago* **13,** 639

SPIELMEYER W. (1908) *Histol. histopathol. Arb.* **2,** 193

TAY W. (1881) *Trans. opthal. Soc. U.K.* **1,** 55

TERRY R.D. and KOREY S.R. (1963) *J. Neuropath. exp. Neurol.* **22,** 98

TERRY R.D. and WEISS M. (1963) *J. Neuropath. exp. Neurol.* **22,** 18

THIEFFRY S. (1962) *Pédiatrie* **17,** 593

VAN BOGAERT L. and SCHOLZ W. (1932) *Zentbl. ges. Neurol. Psychiat.* **141,** 510

VOGT H. (1905) *Mschr. Psychiat. Neurol.* **18,** 161

WILSON S.A.K. (1912) *Brain* **34,** 295

WITTE F. (1921) *Münch. med. Wschr.* **68,** 69

CHAPTER EIGHTEEN

GENETICALLY DETERMINED NEUROLOGICAL DISEASES OF CHILDREN

JOHN N.CUMINGS

INTRODUCTION

The clinical aspects of some of the genetically determined disorders in children have been presented in the preceding chapter; to describe each in detail or to include every possible condition would be impossible in one short chapter. This is particularly true as regards the biochemical features and hence only a limited number of conditions will be discussed and these are listed in Table 1.

TABLE 1

Some genetically determined neurological diseases in children

---

Sphingolipidoses
Refsum's disease
Abetalipoproteinaemia
Hepatolenticular degeneration
Hyperuricaemia
Infantile necrotizing encephalopathy
Huntington's chorea
Friedreich's ataxia
Hereditary fructose intolerance

THE SPHINGOLIPIDOSES

Numerous lipid disorders of the brain occur in childhood and for many decades were classified according to their clinical and histological characteristics. Thus some were described as lipidoses where deposition of some lipid compound was discernible histologically, one characteristic example being Tay-Sachs disease. Some were called leucodystrophies, for the white matter appeared to be mainly the affected area, such was metachromatic leucodystrophy. It was for many years the accepted view that these were cerebral disorders and that the other tissues of the body were not involved. This was demonstrably false in regard to Niemann-Pick's disease and Gaucher's disease, and such authorities as Brain and Greenfield

TABLE 2

Sphingolipid diseases investigated

Condition	Number
Gaucher	4
Niemann-Pick	8
Metachromatic	39
Krabbe	14
Tay-Sachs	12
'Amaurotic family idiocy'	57

TABLE 3

Sphingolipids

Sphingomyelin	Sphingosine + FA + Phosphorylcholine
Cerebroside	Sphingosine + FA + Hexose
Sulphatide	Sphingosine + FA + $\begin{cases} \text{Hexose} \\ \text{SO}_4 \end{cases}$
Ganglioside	Sphingosine + FA + $\begin{cases} \text{Hexose} \\ \text{NANA} \\ \text{Hexosamine} \end{cases}$

Sphingosine + Fatty acid (FA) = Ceramide.

(1950) and Norman (1947) showed that organs other than the brain were involved in some forms of leucodystrophy.

Within the past few years, as a result mainly of electron microscopy (EM) and biochemical studies, many of the conditions previously grouped under these two main headings now more precisely belong to a group in which there is an abnormality of sphingolipid metabolism. Hence the term sphingolipidoses for these conditions, in many of which a specific enzyme defect has been uncovered. Further study will result in a more exact classification of the group of conditions known as amaurotic family idiocy; in fact, some of these cases have already been more accurately delineated.

Table 2 lists the more common of these diseases and the numbers of the cases which have been examined personally. Each of these diseases is the result of an abnormality in a particular sphingolipid and in its metabolic pathway. Each major sphingolipid has sphingosine as a part of its structure as illustrated in Table 3. This is an over-simplification, for the amounts of the fatty acids vary from one lipid to another even though

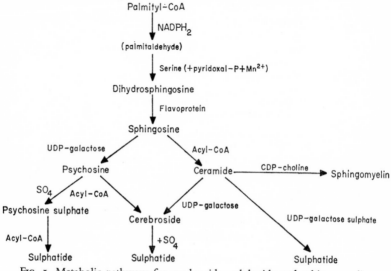

FIG. 1. Metabolic pathways for cerebroside, sulphatide and sphingomyelin.

C18, C20 and C24 are those most frequently found. Further, in the gangliosides differing molar proportions of sphingosine, hexose and *n*-acetyl neuraminic acid (NANA) are present, and in one NANA-containing compound hexosamine is absent. Shortly, mention will be made of each of the diseases and the abnormality present, but the metabolic pathway of one or two compounds may be illustrated in Fig. 1 which shows some of the suggested or proven steps by which cerebroside, cerebroside sulphate or sulphatide and sphingomyelin are formed. The metabolic pathways for gangliosides are also known for the most part, but some details are still required in what is a relatively more elaborate scheme than that for cerebrosides (Svennerholm, 1967). Later, some indication of the enzymic abnormalities in these pathways will be given for some of the diseases, but it is simpler to describe first the composition of the cerebral lipids in abnormal states.

The normal lipid composition of the brain has been discussed frequently in previous papers (Cumings, 1964) and it will suffice to note the changes from the normal as each disease is discussed.

GAUCHER'S DISEASE

It is known that in this condition the organs, such as the spleen, contain abnormally high concentrations of cerebroside. This is also true of the

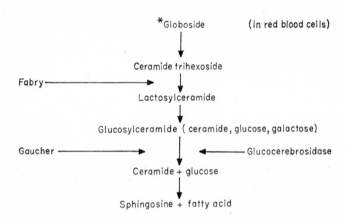

*In the brain ganglioside probably replaces globoside in the chain

FIG. 2. Metabolic pathway to indicate site of enzymic abnormality in Gaucher's disease and in Fabry's disease.

brain, but to a much less extent (Maloney and Cumings, 1960). A thin-layer chromatogram (TLC) of the splenic lipids demonstrates this raised cerebroside level. However, in the normal brain the hexose in cerebroside is galactose, whereas in Gaucher's disease about one-third is glucose. It has been demonstrated that a lack of glucocerebrosidase is responsible for the abnormality present. Fig. 2 shows the metabolic pathways involved and the site of action of the enzyme, from which it can be seen that a block at the site indicated will result in retention of both glucose and galactose in cerebroside.

Fig. 2 also shows the site of another enzymic block which results in a rare disorder known as Fabry's disease; in this disorder there is an accumulation, especially in the kidney, of ceramide trihexoside.

NIEMANN-PICK'S DISEASE

This condition when affecting the brain is somewhat more common than Gaucher's disease. It has been known since 1934 that the deposited lipid, seen microscopically in many tissues, consists essentially of sphingomyelin (Klenk, 1934), and this is readily demonstrable in a TLC of cerebral lipid extracts. In addition, the organs show raised levels of cholesterol, while in the brain gangliosides are raised. When examined by TLC two additional well-marked bands known as G_{M3} and G_{M4} have been found (see Booth, Goodwin and Cumings, 1966). The reason for the last two abnormalities is unknown, but it has been demonstrated by Brady (1967) that the enzyme sphingomyelinase is very greatly reduced in Niemann-Pick's disease in the tissues.

It is of interest that Brady (1967) has made a study of the blood leucocytes and employing the incorporation of radioactive isotopes has been able to demonstrate an absence or gross reduction in each of the three appropriate enzymes in the three conditions so far mentioned.

METACHROMATIC LEUCODYSTROPHY

Rarely this condition affects adults as well as children, but biochemically the findings are similar. As with those diseases already described, the distribution of the lesions is widespread although the areas involved apart from the brain are usually the kidney, peripheral nerves and gall-bladder. The main abnormality is an increase in sulphatide in the brain and other organs involved, as demonstrated originally by Austin (1959) and since confirmed in all our more recent cases. Similar results have been obtained by me from peripheral nerves and from the kidney.

Sulphatide is a myelin lipid, and when pure myelin is obtained by a suitable technique the various sphingolipids can be estimated, and this has been done both in brain tissue from normal controls and in metachromatic leucodystrophy. Normal myelin contains no gangliosides and four times as much cerebroside as sulphatide, whereas metachromatic myelin does show a little ganglioside and twenty times as much sulphatide as cerebroside. This latter myelin does not appear to be entirely similar to normal myelin as can be seen from its EM appearance, for small fragments only of typical myelin remain (Cumings, Thompson and Goodwin, 1968).

The enzymic abnormality has been studied by Austin, Armstrong and Shearer (1965) and by Mehl and Jatkewitz (1965) and their colleagues. They have demonstrated that there is a generalized absence of cerebroside sulphatase, and the latter workers have shown this to be true in some of the cases seen here. It is of interest that this enzyme defect can be detected by an examination of the urine of the patient. In addition, there is an accumulation of metachromatic material within the cellular elements in the urine, and our cases have also shown a raised level of sulphatide in the urine.

KRABBE'S DISEASE

This is a relatively rare disorder, but as with metachromatic leuco-dystrophy more than one sib in a family is frequently affected. Further, Bachhawat, Austin and Armstrong (1967) have provided evidence of an enzymic abnormality in the tissues. The enzyme which is reduced or even absent is cerebroside sulphotransferase, which normally is concerned with the acceptance by cerebroside of the sulphate radicle. The abnormalities which can be found by a pathologist are firstly a raised protein content in the CSF, usually a figure of over 100 mg per 100 ml. Other forms of diffuse sclerosis at a similar age incidence do not show such an elevation of protein.

Alterations in lipid content in the brain are not remarkable, but there is evidence that sulphatides are diminished as compared with cerebroside levels so that the cerebroside–sulphatide ratio is now greater than normal. This can be demonstrated by preparing myelin and estimating sphingo-lipids, when it is seen that the ratio of cerebroside to sulphatide is 7:1 (Cumings *et al*, 1968).

There is one other feature which is detected when a lipid extract is examined by TLC for gangliosides. The two bands of G_{M3} and G_{M4} are present, similar in position to those seen in Niemann-Pick's disease (Fig. 3).

TAY-SACHS DISEASE

There are two major biochemical features, one abnormality being detectable in the blood and the other in nervous tissues. Aronson, Perle, Saifer and Volk (1962) found that, of the many enzyme systems in blood and CSF which they examined, one was virtually absent in patients with Tay-Sachs disease. This enzyme, fructose-1-phosphate aldolase, is normally present in from 0.3 to 1.1 FPA units per ml in the serum but in some twenty-four cases of Tay-Sachs disease examined, none contained this enzyme. We have only studied a very few such cases, but with close agreement, for there has been a very gross diminution even in non-Jewish patients. The American workers also showed a considerable reduction of the enzyme levels in the blood of the parents, and to a lesser extent in some of the grandparents.

The abnormality in the brain and at least in some peripheral nerves is a raised concentration of ganglioside, first shown by Klenk (1942), as well as in an abnormal ganglioside pattern as seen in TLC demonstrated by Müldner, Wherrett and Cumings (1962) and by Svennerholm (1962). This abnormality is the presence of an abnormal band G_{M2} which as shown by Wherrett and Cumings (1963) contained some 80–85 per cent of the total NANA present. Recently Cumings *et al* (1968) have been able to demonstrate that normal microsomes contain ganglioside, while

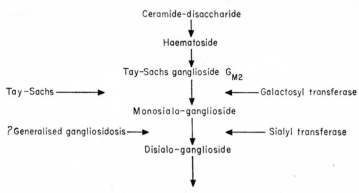

Fig. 4. Metabolic pathway to indicate site of enzymic abnormality in Tay-Sachs disease and generalized gangliosidoses.

those from the cortex of a case of Tay-Sachs disease possess a considerably increased amount, as well as showing the characteristic pattern. This G_{M2} ganglioside has also been found by us to be present in the sciatic nerve.

It has been suggested that the enzyme defect in this condition is an absence of galactosyl transferase which normally acts at the stage where an additional galactose is added to G_{M2}, transforming it to G_{M1} with its three hexose moles as compared to the two in G_{M2}. Details as to the relative degree in deficiency in this enzyme are not yet available. Fig. 4 indicates the site.

AMAUROTIC FAMILY IDIOCY

Clinicians and pathologists have for many years described a variety of conditions under this title, and have tended to distinguish them according to age groups, to the presence of optic atrophy and to variations in special staining techniques in histological preparations. Specific biochemical changes appeared to be lacking, although there was a loss of most lipids

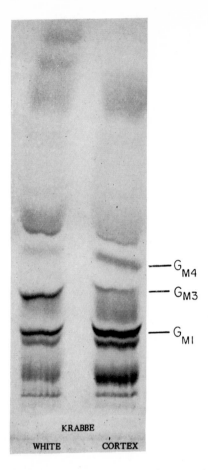

Fɪɢ. 3. Thin-layer chromatogram to show ganglioside pattern in lipid extracts from the brain in Krabbe's disease. Note bands G_{M3} and G_{M4}.

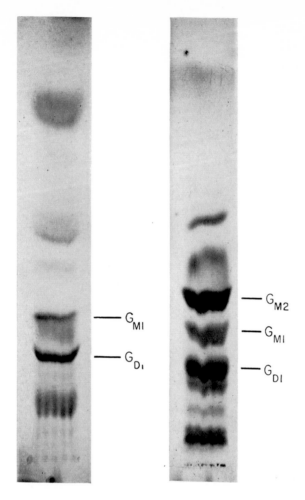

FIG. 5.   FIG. 6.

FIG. 5. Thin-layer chromatogram to show ganglioside pattern in a case of 'amaurotic family idiocy'. Note raised level of G_{D1} and lowered G_{M1}. The patient was a boy of $2\frac{1}{2}$ years.

FIG. 6. As Fig. 5 in another case showing raised G_{M2} and G_{D1} but with a slightly decreased G_{M1}. The patient was a boy of $5\frac{2}{3}$ years.

and in long-standing cases, the presence of cholesterol esters indicating demyelination. Gangliosides did not appear increased, although by comparison with other lipids were not as markedly reduced. Recently some of these cases have been separated from others partly on EM appearances and partly by changes in ganglioside pattern on TLC. It was further found that the ganglioside abnormalities were to be detected in organs other than the brain and the term generalized gangliosidoses has been used (Landing, Silverman, Craig, Jacoby, Lahey and Chadwick, 1964; O'Brien, Stern, Landing, O'Brien and Donnell, 1965). In cases such as those described by Landing *et al*, Brady (1967) has suggested that the enzyme sialyl transferase is reduced so that the block results in an increase of monosialoganglioside or G_{M1}, as indicated in Fig. 4.

We have now examined a number of cases showing unusual patterns. Two showed a raised G_{D1} and lowered G_{M1} (Fig. 5), one a raised G_{M1} and G_{M2} and a reduced G_{D1}, while another revealed a raised G_{M2} as well as G_{D1}, but with a slightly decreased G_{M1} (Fig. 6).

There is obviously much more to be learnt about this group of conditions before we can properly classify them.

REFSUM'S DISEASE

This condition, while it is more commonly recognized in the adult, may commence in youth. Until recently the only abnormality found with any regularity was a raised CSF protein, the level of which fluctuated from time to time in the course of the disease. Refsum (1946), who originally described the condition, regarded it as similar to the lipidoses and Cammermeyer (1956) supported this view. In a recent article Refsum (1967) claims it is a disease in its own right and one of the group of inborn errors of metabolism.

Klenk and Kahlke (1963), together with Richterick, Kahlke, van Mechelen and Rossi (1963), discovered that there was an excess of phytanic acid (3, 7, 11, 15-tetramethyl hexadeconic acid) in blood and in some tissues such as the liver and kidney. In the past 3 years numerous workers have recorded cases with similar findings (see Nevin, Cumings and McKeown, 1967). Phytanic acid has been found by us to be present in muscle, fat, nerve, liver, blood, urine and in one CSF, in patients examined here.

A number of other examinations have been made, but in general it can be said that none of the cases examined has shown abnormal copper, caeruloplasmin or creatine kinase levels in the sera.

T

The abnormality appears to be the result of an inborn defect in a degradation pathway involving certain branched-chain fatty acids. Eldjarn, Try and Stokke (1966) found that a pathway involving a CO_2-fixation mechanism was lacking in subjects with Refsum's disease. This alternative degradation pathway differs from β-oxidation, ω-oxidation and the isovaleric degradation pathways. They used ^{14}C-3,6-dimethylcaprylic acid to demonstrate this pathway. Stokke, Try and Eldjarn (1967) have recently published further observations. The conclusion they draw is that in patients with Refsum's disease there is a failure to degrade β-methyl fatty acids by a decarboxylation process. This they demonstrated by injection of 3,6-dimethyl [8-^{14}C] octanoic acid. Normal subjects excreted $^{14}CO_2$ through the lungs and a metabolite 2,5-dimethyl-heptanoic acid in the urine, whereas in the patients completely negative results were obtained.

The same group of workers have attempted to alter this abnormality by means of a dietary restriction of phytanic-acid precursors (Eldjarn, Try, Stokke, Munthe-Kass, Refsum, Steinberg, Avigan and Mize, 1966). Thus foods with a high content of phytol have been omitted from the diet. The results have been variable, some patients improving, some showing no change. However, remissions and relapses take place without therapy so that a striking improvement would be needed if such therapy were to be regarded as successful. One of the cases whose blood I have examined has not shown any really significant variation during therapy.

ABETALIPOPROTEINAEMIA

This rare but most interesting condition has been well reviewed recently by a number of workers (Wolff, 1965; Farquhar and Ways, 1966). As the name implies, there is an absence or very gross diminution of β-lipoprotein, but it is important to note that the high density—1.063 to 1.21—lipoproteins are also reduced but not to the same degree as the low-density ones. As there is an intestinal defect, there is an absence in the blood of chylomicrons postprandially. Additionally there is a peculiar change to be seen in the red cells; which cells are then known as acanthocytes; this alteration in shape is found in wet preparations. Associated with an absence of the low-density lipoproteins there are cholesterol and phospholipid abnormalities.

Red blood cells

The appearance is that of crenated cells with pseudopodic processes. Normal red blood cells mixed with serum from a patient are not changed, but when normal red blood cells are infused into a patient they are altered. It is possible by the addition of detergents such as Tween 80, containing unsaturated but not saturated long-chain fatty acids, to acanthocytes to cause them to return to a normal shape (Switzer and Eder, 1962). The red cells in a patient which are young are normal, bone marrow cells being unaffected, whereas the degree of deformity increased with age (Simon and Ways, 1964). There is a slightly increased sphingomyelin content but the linoleic acid (18:2) content of phospholipids and of cholesterol is reduced to about 75 per cent of the normal (Ways, Reed and Hanahan, 1963).

Serum lipoprotein
The low-density (<1.063) β-lipoprotein is virtually absent in this disease; the α-lipoprotein is diminished, for it has been suggested that this is derived from β-lipoprotein; however, low levels of α-lipoprotein are found in other forms of defective intestinal absorption. Lees (1967) in a recent paper, as a result of certain immunological studies, has suggested that both normal subjects and those with this genetically determined disease possess a lipid-free apoprotein of β-lipoprotein named by him B protein. The disease is thus not due to an inability to synthesise the apoprotein but is a defect in the formation of the β-lipoprotein macro-molecule. The complete lack of glyceride transport in the disease suggests that β-lipoprotein is essential for the process. The presence of soluble B protein in plasma would be consistent with a transport phenomenon at the intestinal site of utilization where combination of lipid and B protein might take place.

Phospholipids and cholesterol
The serum levels of phospholipids usually range from about 20 to 95, cholesterol from 20 to 70 and triglycerides 10 to 15 all in mg per 100 ml, all of which are considerably reduced below the normal. The changes in linoleic acid in serum are similar to those in red cells.

There is a moderate to severe steatorrhoea, the result of the intestinal defect.

Almost always the heterozygotes are both symptom free and without definite biochemical abnormalities.

WILSON'S DISEASE

This condition in children may exactly resemble that seen in older subjects although this is more commonly true of children over the age of 10 years.

The biochemical variations from the normal are well known and descriptions can be found elsewhere (Cumings, 1959, 1968; Walshe, 1967). However, there are a few different modes of presentation discernible in the young subject.

Sjövall and Wallgren in 1934 and Brinton (1947) described haemolytic crises with jaundice in young patients, the former's case being a child of 10 years. The degree of haemolytic anaemia varies, as does the jaundice. Usually neurological signs are minimal or even absent, and in almost all of the patients described the diagnosis of Wilson's disease was not made until there was recovery from the haemolytic attack, which may be quite transient. Very occasionally evidence of Wilson's disease is present before the haemolytic episode. McIntyre, Clink, Levi, Cumings and Sherlock (1967) have described three such cases in whom investigations were made to discover the cause of the haemolysis. Serum erythrocyte and urine coppers, caeruloplasmin levels as well as red cell half-life using ^{51}Cr were made. In two of the three cases the copper content in the liver was also estimated. The observations confirmed the diagnosis of Wilson's disease for caeruloplasmin levels were diminished and urinary copper was raised above the normal. Further, there was evidence to suggest that a flux of copper took place during the period of haemolysis, but this was shown not to be the result of the haemolysis but probably to be derived from the tissues by liberation of copper. It is therefore probable that this sudden release of tissue copper results in haemolysis of the red blood cells. It should be added that in two of the patients the liver copper was raised, in one to 87 and in the other to 162 mg per 100 g dry tissue, compared to a normal level of under 20 mg per 100 g.

Another form of presentation is the child whose first symptom is jaundice, the result of an early cirrhosis when sometimes the correct diagnosis is not revealed even at autopsy; but a sib with similar signs coming to hospital for investigation, the possibility of Wilson's disease is then considered. Chambers, Iber and Uzman in 1957 discussed the features found in such cases, and Sass-Kortsak, Glatt, Cherniak and Cederlund (1961) reported further in similarly affected children. These last authors found that in a 10 year period there were fifty cases of cirrhosis in patients under 15 years of age. A reinvestigation of eighteen cases of apparently idiopathic post-necrotic cirrhosis led to the finding of five cases of Wilson's disease, in none of whom neurological signs or symptoms were present, yet these patients showed typical biochemical features of that disease. This has been our experience, for out of a total of eighty-five cases of Wilson's disease examined, seventy-seven showed a Kayser-Fleischer ring, and of the eight others this sign was not sought for in three, while the other five were all in non-symptomatic children with liver cirrhosis (Cumings, 1968). This, then, is a condition in which Wilson's disease must be considered as a possibility. Such children almost always reveal copper abnormalities in blood and urine of the type seen in the adult case. They respond very well to therapy.

It should be mentioned that the diagnosis of Wilson's disease in asymptomatic sibs of an affected patient in whom even cirrhosis is not suspected can be very difficult. This problem has been discussed recently

TABLE 4

Liver copper content: specimens obtained by needle biopsy in mg per 100 g dry tissue

Number	Group	Results
32	Normal or control subjects	0.87 to 19.5
3	Unaffected sibs of Wilson's disease subjects	3.16 to 18.6
9	Wilson's disease 6 with neurological symptoms 3 asymptomatic and sibs of known cases	35.5 to 74.8

by Levi, Sherlock, Scheuer and Cumings (1967) who found that a liver needle biopsy with histological and biochemical studies is a valuable tool in diagnosis. It was reported that the normal liver content, using this technique, is under 20 mg per 100 g dry tissue. A copper content of over 30 mg per 100 g of dry tissue was regarded as indicating Wilson's disease and treatment can be commenced. Table 4 indicates the range of figures found and when raised levels of liver copper are present, therapy with penicillamine should be commenced even in the absence of neurological signs.

HYPERURICAEMIA

Since 1959 a number of papers have recorded an unusual clinical syndrome involving boys in which self-mutilation and a raised blood uric acid are prominent. The condition was well described in 1964 by Lesch and Nyhan although the disease had almost certainly been reported before in papers from Germany and England. However, the American authors found a raised blood content and an increased urinary excretion of uric acid. Lesch and Nyhan demonstrated that when glycine was administered the formation of uric acid exceeded by up to 200 times the normal. Sass, Itabashi and Dexter (1965) found changes in the brain of a fatal case like those of uraemia together with some possible deposits of urate crystals in the cerebellum. Hoefnagel, Andrew, Mireault and Berndt (1965) say that, assuming the condition to be sex-linked then typing for the Xg blood group indicated that the crossover must have occurred between the Xg locus and the locus for the disorder.

All the six cases described by Michener (1967) showed the abnormal biochemical features. The serum uric acid was up in all, usually over 10 mg per 100 ml. One case examined here was over 20 mg per 100 ml. Two out of three of Michener's cases had a raised CSF level of uric acid. The urinary output of uric acid in the normal control ranges from 7.7 to 17.7 mg per kg body weight, but in these six boys it ranged from 13.1 to 66.5 mg per kg. These patients were treated with probenecid (Benemid) with reduction of serum levels, but clinically there was no significant improvement. On the next admission of one boy fluids intravenously were as effective as the drug in reducing the serum uric acid.

Dodge (1966) recently pointed out that all cases show these abnormal biochemical findings with blood uric acid ranging from 9 to 25 mg per 100 ml, while Partington and Hennen (1967) described two similar cases who responded to allopurinol. Lesch and Nyhan suggested that there is an incompletely developed blood-brain barrier in these boys so that large quantities of uric acid enter the brain and this results in damage. This is a little akin to the basal ganglia damage in kernicterus associated with a raised level of circulating bilirubin of over 20 mg per 100 ml.

However, further studies have yielded some clues as to the basic biochemical abnormality. In 1966 Kaplan, Wallace and Halberstam, after experiments with glycine, demonstrated an increased uptake of glycine by the renal tubule cells. They postulated that there was at least a local cell membrane defect, which could be universal throughout the body, so that there was both a renal retention of uric acid and an over production because of an increased availability of purine precursor and increased purine production.

Seegmiller, Rosenbloom and Kelley (1967) conducted a series of most interesting experiments. First they tried the effect of oxathiprine on purine metabolism using glycine 1-^{14}C. Purine synthesis was depressed in gouty patients but not in the affected children. Then they tested the effect of 6-mercaptopurine. It is said that resistance to the action of this compound is associated with a deficiency of a transferase enzyme acting on hypoxanthine—guanine phosphoribosyl transferase (Fig. 7). They tested the effects on tissue culture monolayers of fibroblasts. 6-Mercaptopurine inhibited purine biosynthesis in normal cells but did not do so in cells from affected children. Similar results were obtained using red blood cells; substances A and B were reduced by this method provided suitable substrates were used. The problem that remains is exactly where it works and why uric acid is increased. Normally the absence of an enzyme results in a lack of a compound, not an increase. This would suggest that a regulatory mechanism may also be involved.

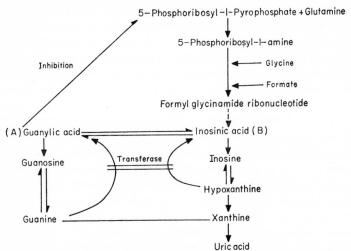

FIG. 7. Some pathways in purine metabolism.

Perhaps one might add that the patient seen in this hospital was operated upon by Mr Valentine Logue in order to suppress the athetotic movements and for a short period of time this was successful, but recurrence of the movements took place. It may further be added that at present few adequate examinations of the brain or other organs have been made. One small clinical point is that uric-acid stones may be found, but these can usually be eliminated by the use of alkalies, but this does not alter the neurological findings.

INFANTILE NECROTIZING ENCEPHALOPATHY

Leigh (1951) described the clinical and pathological features of a 7-month-old boy and gave the condition the name of infantile necrotizing encephalopathy. The subacute necrotic lesions, mainly in the central grey matter, were symmetrical and were thought to resemble lesions in monkeys following administration of 8-aminoquinolines. Denny-Brown, however, considered that there was a close resemblance to Wernicke's encephalopathy and suggested that the lesions were the result of either thiamine deficiency or of an anti-thiamine effect.

Since then a number of cases have been described, all with similar clinical and histological features. Feigin and Wolf (1954) and Ule (1959) although agreeing in general with Denny-Brown, did not find evidence for B_1 deficiency, although the former authors suggested an enzyme deficiency of a congenital nature. Ebels, Blokzijl and Troelstra (1965) found the condition in children up to the age of 15. All consider there is an inborn error of thiamine deficiency. Worsley, Brookfield, Elwood, Noble and Taylor (1965) noted that hyperpnoea and muscle twitching were present and they performed interesting biochemical studies, for they found a spontaneous increase in blood lactic acid. They showed that the patients' red cells formed lactic acid by glycolysis faster and in a greater amount (17 per cent) than normal red cells, and suggested that the proportion of phosphorylated hexoses to free glucose in the blood was raised. They obtained levels of blood pyruvate of 2.1 and 2.4 mg per 100 ml with a lactate level of 54 per 100 ml. There was also a renal aminoaciduria and a lowered phosphate concentration. One practical point emerged in that estimation of true glucose values in the blood were necessary for otherwise phosphorylated hexoses and hexose phosphates would be estimated and an apparently normal blood sugar found.

Kakke, Ebels and ten Thye (1967) have recorded further examples and listed the thirty-two cases so far described which included six juvenile patients. In their cases the addition to the diet of B complex did not affect the course of the disease.

More recently, Procopis, Turner and Selby (1967) have discussed another case in an acidotic child with special reference to the possible causes of the cerebral lesions. No firm conclusions were revealed or indeed seem possible in the present state of knowledge.

Clayton, Dobbs and Patrick (1967) from Great Ormond Street have investigated three further cases. They have specially studied changes in pyruvate and citrate in blood and urine as well as various enzymes in liver, brain and muscle. A raised blood pyruvate was regularly found, a slightly low serum citrate and raised urinary citrate less frequently observed. The only major enzymic abnormality was a low muscle phospho-fructokinase. No specific defect could be determined and the nature of the disease has not been elucidated. However, therapy with lipoate was used on the assumption that this might enable intermediate steps in the oxidation of pyruvate to be accelerated and so pyruvate would be metabolized more readily.

HUNTINGTON'S CHOREA

Biochemical studies have been unrewarding thus far though a biochemical lesion is probable even in young children such as those described by Byers and Dodge (1967).

Kenyon and Hardy (1963) found raised levels of magnesium and calcium in the red blood cells although the serum concentrations were normal. This work was repeated together with the use of isotopically labelled ions by Bruyn, Mink and Caljé (1965) and by Jones, Desper and Flink (1965) but no abnormalities were detected by these workers or by Fleming, Barker and Stewart (1967). No abnormality in serum caeruloplasmin or copper, in urinary amino acids, indoles or dopamine have been found and the CSF is normal.

Courville, Nusbaum and Butt (1963), during their examination for trace metals in brains from many diseases, found that in three cases of Huntington's chorea levels of strontium were raised, ranging it seemed from 0.1 to 0.2 mg per 100 g dry tissue, while the copper content was diminished in amount. In two cases we have examined by a crude technique there was slightly more strontium present than normally but the amounts present were not significantly raised, but the copper and zinc levels were not altered from the normal.

It is possible that an approach to cellular metabolic changes would be more profitable.

FRIEDREICH'S ATAXIA

The cause of this disease, which also appears to be an inborn error of metabolism, is still obscure. The findings thus far are that there appears to be a higher incidence of a diabetic type of blood-sugar curve than in control groups of subjects—thus in one group of sixty-five patients there were eleven, and in another of fifty patients there were nine such abnormal curves. Enzyme studies of cerebral tissue have suggested that there is an alteration from a glycolytic to a direct oxidative route in carbohydrate metabolism (Robinson, 1966a). The blood enzymes (Robinson, Phillips and Cumings, 1965) and the urinary amino acids and indoles are not altered (Robinson, Curzon and Theaker, 1965).

A myocardiopathy is frequently present (30–55 per cent of cases) but the only enzyme abnormality found has been a reduction in aldolase (Robinson, 1966b).

As in Huntington's chorea dynamic studies of cell metabolic processes are likely to be of greater value than the more simple biochemistry available thus far.

HEREDITARY FRUCTOSE INTOLERANCE

Fructose intolerance as a genetically determined condition was first described in 1956 by Chambers and Pratt at the National Hospital. A year later Froesch, Prader, Labhart, Stuber and Wolf (1957) from Switzerland showed the biochemical responses in more detail to a dose of fructose in affected individuals. Some forty cases have now been described and five such children have recently been reported upon by Black and Simpson (1967).

Young infants, after being changed from breast feeding to dried milk with added sucrose, vomit, with failure to thrive and often a subsequent jaundice. If glucose replaces the sucrose, normal growth takes place, but this alteration in the diet is only possible after the recognition of the nature of the disorder.

The diagnosis is readily confirmed by administration of a dose of fructose by mouth—usually 0.5 g per kg body weight—or alternatively by intravenous injection. Blood levels of glucose, fructose and phosphate are measured and it is usually found that the blood sugar falls, with clinical evidence of hypoglycaemia in some cases. The level of fructose varies, usually rising and speedily falling, but there is no constant pattern (Cornblath, Rosenthal, Reisner, Wybregt and Crane, 1963). Inorganic phosphate levels fall often before the drop in blood glucose occurs.

After fructose by mouth there may be an aminoaciduria as well as traces of a reducing substance, which by chromatography can be shown to be fructose as well as glucose. This, in fact, was true of the case described by Chambers and Pratt, where amounts of fructose in the urine varied with the dose and type of sugar given. It must be mentioned that the use of glucose specific test papers are valueless for the urine examination of these patients.

The primary enzyme defect in hereditary fructose intolerance is a lack of a specific aldolase in the liver (see Fig. 8). Assays of fructose-1-phosphate splitting aldolase show a gross reduction to between zero and 12 per cent of the normal (Nikkila, Somersalo, Pitkanen and Perheentupa, 1962).

Another enzyme fructose-1-6-diphosphate splitting liver aldolase is also reduced but only to about half of its normal value. Normally the ratio of these two enzymes is as 1 to 1, but it is greater than 6 to 1 in affected individuals. Liver fructokinase is normal, whereas it is lacking in essential fructosuria. These abnormalities can be detected in the liver both by biochemical methods and by histochemical examination (Lake, 1965).

The explanation for the hypoglycaemia after fructose is still not entirely resolved. There may be a block in the release of glucose from the liver; severe hypoglycaemia appears to follow a failure of glycogenolysis in the liver.

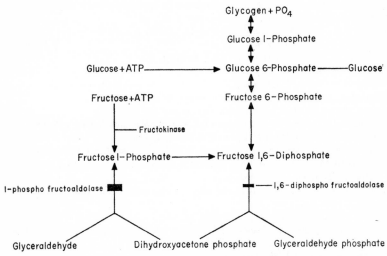

FIG. 8. Metabolic pathways to indicate site of block in hereditary fructose intolerance.

However, if galactose is given with fructose to the patients, hypo-glycaemia is less marked and is of shorter duration (Cornblath *et al*, 1963). One cause of diminished glycogenolysis might be a lack of inorganic phosphate within the liver cell, reflected by the lower blood levels.

Treatment consists of the removal from the diet of fructose and sucrose containing foods when normal growth in children is possible and prognosis is very good.

CONCLUSION

It would appear to be highly probable that when this subject of neuro-logical diseases in children is discussed after another decade many new and important disorders will be described in which defects of metabolism will feature prominently. Further, that of those mentioned in this chapter the specific aetiological factor in each will then be known and appropriate therapeutic procedures will have been discovered.

REFERENCES

Aronson S.M., Perle G., Saifer A. and Volk B.W. (1962) *Proc. Soc. exp. Biol. N.Y.* **111**, 664

Austin J.H. (1959) *Proc. Soc. exp. Biol. N.Y.* **100**, 361

Austin J.H., Armstrong D. and Shearer L. (1965) *Arch. Neurol. (Chicago)* **13**, 593

Bachhawat B.K., Austin J. and Armstrong D. (1967) *Biochem. J.* **104**, 15C

Black J.A. and Simpson K. (1967) *Brit. med. J.* **ii**, 138

Booth D.A., Goodwin H. and Cumings J.N. (1966) *J. Lipid Res.* **7**, 337

Brady R.O. (1967) *Clin. Chem.* **13**, 565

Brain W.R. and Greenfield J.G. (1950) *Brain* **73**, 291

Brinton D. (1947) *Proc. Roy. Soc. Med.* **40**, 556

Bruyn G.W., Mink C.J.K. and Caljé J.F. (1965) *Neurology (Minneap.)* **15**, 455

Byers R.K. and Dodge J.A. (1967) *Neurology (Minneap.)* **17**, 587

Cammermeyer J. (1956) *J. Neuropath. exp. Neurol.* **15**, 340

Chalmers T.C., Iber F.L. and Uzman L.L. (1957) *New Engl. J. Med.* **256**, 325

Chambers R.A. and Pratt R.T.C. (1956) *Lancet* **ii**, 340

Clayton B.E., Dobbs R.H. and Patrick A.D. (1967) *Archs Dis. Childh.* **42**, 467

Cornblath M., Rosenthal I.M., Reisner S.H., Wybregt S.H. and Crane R.K. (1963) *New Engl. J. Med.* **269**, 1271

Courville C.B., Nusbaum R.E. and Butt E.M. (1963) *Arch. Neurol. (Chicago)* **8**, 481

Cumings J.N. (1959) *Proc. Roy. Soc. Med.* **52**, 61

Cumings J.N. (1964) In *Diseases of Metabolism*, 5th ed. p. 1405. Ed. Duncan G.G. W.B.Saunders Company, Philadelphia

Cumings J.N. (1968) *J. clin. Path.* **21**, 1

Cumings J.N., Thompson E.J. and Goodwin H. (1968) *J. Neurochem.* **15**, 243

Dodge P.R. (1966) *Develop. Med. Child Neurol.* **8**, 89

Ebels E.J., Blokzijl E.J. and Troelstra J.A. (1965) *Helv. Paed. Acta* **20**, 310

Eldjarn L., Try K. and Stokke O. (1966) *Biochim. Biophys. Acta* **116**, 395

Eldjarn L., Try K., Stokke O., Munthe-Kass A.W., Refsum S., Steinberg D., Avigan J. and Mize C. (1966) *Lancet* **i**, 691

Farquhar J.W. and Ways P. (1966) In *The Metabolic Basis of Inherited Disease*, 2nd ed. p. 509. Ed. Stanbury J.B., Wyngaarden J.B. and Fredrickson D.S. McGraw-Hill Book Co., New York

Feigin I. and Wolf A. (1954) *J. Pediat.* **45**, 243

Fleming L.W., Barker M.G. and Stewart W.K. (1967) *J. Neurol. Neurosurg. Psychiat.* **30**, 374

Froesch E.R., Prader A., Labhart A., Stuber H.W. and Wolf H.P. (1957) *Schweiz med. Wschr.* **87**, 1168

Hoefnagel D., Andrew E.D., Mireault N.G. and Berndt W.O. (1965) *New Engl. J. Med.* **273**, 130

JONES J.E., DESPER P.C. and FLINK E.B. (1965) *Metabolism* 14, 813
KAKKE J.P.W.F., EBELS E.J. and TEN THYE O.J. (1967) *Arch. Neurol. (Chicago)* 16, 227
KAPLAN D., WALLACE S.L. and HALBERSTAM D. (1966) *Nature (Lond.)* 209, 213
KENYON F.E. and HARDY S.M. (1963) *J. Neurol. Neurosurg. Psychiat.* 16, 123
KLENK E. (1934) *Hoppe Seyl. Z.* 229, 151
KLENK E. (1942) *Ber. dt. chem. Ges.* 75, 1632
KLENK E. and KAHLKE W. (1963) *Hoppe Seyl. Z.* 333, 133
LAKE B.D. (1965) *Jl R. microsc. Soc.* 84, 489
LANDING B.H., SILVERMAN F.N., CRAIG J.M., JACOBY M.D., LAHBY M.E. and CHAD-
    WICK D.L. (1964) *Am. J. Dis. Child.* 108, 503
LEES R.S. (1967) *J. Lipid Res.* 8, 396
LEIGH D. (1951) *J. Neurol. Neurosurg. Psychiat.* 14, 216
LESCH M. and NYHAN W.L. (1964) *Am. J. Med.* 36, 561
LEVI A.J., SHERLOCK S., SCHEUER P.J. and CUMINGS J.N. (1967) *Lancet* ii, 575
MALONEY A.F.J. and CUMINGS J.N. (1960) *J. Neurol. Neurosurg. Psychiat.* 23, 207
McINTYRE N., CLINK H.M., LEVI A.J., CUMINGS J.N. and SHERLOCK S. (1967) *New
    Engl. J. Med.* 276, 439
MEHL E. and JATZKEWITZ J. (1965) *Biochem. biophys. Res. Commun.* 19, 407
MICHENER W.M. (1967) *Am. J. Dis. Child.* 113, 195
MÜLDNER H.G., WHERRETT J.R. and CUMINGS J.N. (1962) *J. Neurochem.* 9, 607
NEVIN N.C., CUMINGS J.N. and McKEOWN F. (1967) *Brain* 90, 419
NIKKILA E.A., SOMERSALO O., PITKANEN E. and PERHEENTUPA J. (1962) *Metabolism* 11,
    727
NORMAN R.M. (1947) *Brain* 70, 234
O'BRIEN J.S., STERN M.B., LANDING B.H., O'BRIEN J.K. and DONNELL G.N. (1965)
    Am. J. Dis. Child. 109, 338
PARTINGTON M.W. and HENNEN B.K.E. (1967) *Develop. Med. Child Neurol.* 9, 563
PROCOPIS P.G., TURNER B. and SELBY G. (1967) *J. Neurol. Neurosurg. Psychiat.* 30, 349
REFSUM S. (1946) *Acta psychiat. scand.* Suppl. 38
REFSUM S. (1967) *T. norske Laegeforen.* 87, 445
RICHTERICH R., KAHLKE W., VAN MECHELEN P. and ROSSI E. (1963) *Klin. Wschr.* 41,
    800
ROBINSON N. (1966a) *Arch. neuropath. (Berl.)* 6, 25
ROBINSON N. (1966b) *Neurology (Minneap.)* 16, 1135
ROBINSON N., CURZON G. and THEAKER P. (1965) *J. clin. Path.* 18, 797
ROBINSON N., PHILLIPS B.M. and CUMINGS J.N. (1965) *Brain* 88, 131
SASS J.K., ITABASHI H.H. and DEXTER R.A. (1965) *Arch. Neurol. (Chicago)* 13, 639
SASS-KORTSAK A., GLATT B.S., CHERNIAK M. and CEDERLUND I. (1961) In *Wilson's
    Disease. Some Current Concepts*, p. 151. Ed. WALSHE J.M. and CUMINGS J.N.
    Blackwell Scientific Publications, Oxford
SEEGMILLER J.E., ROSENBLOOM F.M. and KELLEY W.N. (1967) *Science* 155, 1682
SIMON E.R. and WAYS P. (1964) *J. clin. Invest.* 43, 1311
SJÖVALL E. and WALLGREN A. (1934) *Arch. psychiat. neurol.* 90, 435
STOKKE O., TRY K. and ELDJARN L. (1967) *Biochim. Biophys. Acta* 144, 271
SVENNERHOLM L. (1962) *Biochem. biophys. Res. Commun.* 9, 436
SVENNERHOLM L. (1967) In *Inborn Errors of Sphingolipid Metabolism*, p. 169. Ed.
    ARONSON S.M. and VOLK B.W. Pergamon Press, Oxford
SWITZER S. and EDER H.A. (1962) *J. clin. Invest.* 41, 1404

ULE G. (1959) *Virchows Arch. path. Anat. Physiol.* **332,** 204

WALSHE J.M. (1967) *Brain* **90,** 747

WAYS P., REED C.F. and HANAHAN D. (1963) *J. clin. Invest.* **42,** 1248

WHERRETT J.R. and CUMINGS J.N. (1963) *Trans. Amer. Neurol. Ass.* **88,** 108

WOLFF H.O. (1965) In *Ergebnisse der Inneren Medizin und Kinderheilkunde*, p. 190. Ed. HEILMEYER L., SCHOEN R. and PRADER A. Springer-Verlag, Berlin

WORSLEY H.E., BROOKFIELD R.W., ELWOOD J.S., NOBLE R.L. and TAYLOR W.H. (1965) *Archs Dis. Childh.* **40,** 492

INDEX

Abetalipoproteinaemia 261, 276–7
Abiotrophies 251, 252, 259–60
Abscess, cerebral, radioisotope uptake and 206
Acanthocytes 261, 276
Acetylcholine, autoantibody muscle receptor blockage and 6
N-acetylphenylalanine 110
Achondroplasia 219
Acromegaly 195
ACTH
    antirabies vaccination and 25–6
    antismallpox vaccination and 29
    in 'autoimmune' demyelinating disorders 40–1
    in multiple sclerosis 39
    memory and 172
Actinomycin, m-RNA synthesis block 171
Addison's disease, lymphorrhages and 6
Adenomas
    basophilic 195
    chromophobe 194–5
    eosinophilic 195
Adenosine diphosphate, cerebral infarction and 152
Adrenal cortex, hyperplasia 196
Adrenaline, blood levels in phenylketonuria 115
AGG see Gamma-globulin, anti-vaccinia
Akathisia 75, 76
Albumin, human serum, I_{131} labelled 221–3
Allo-purinol 263
Alzheimer's disease 223

Amaurotic familial idiocy 156, 254–6 270, 274–5
Amino-acids, inborn metabolic errors
    neurological disorders and 99–128
    biochemical aspects 109–28
    symptoms and signs 101
    table of 100–1
Aminoaciduria
    infantile necrotizing encephalopathy and 282
    mental retardation and 99
γ-aminobutyric acid deficiency, leucinosis and 118
Amitryptiline 85
    brain amine metabolism and 92
    demethylation 88
    effect on surface activity and membranes 87
Ammonia, conversion to urea 121, 122
Amnesia, retrograde 158–9, 161
    temporal lobe stimulation and 162
Amyelination, phenylketonuria and 114–15
Anaemia
    cellular radiosensitivity and 174
    haemolytic
        acquired 44
        Wilson's disease and 278
Aneurysms see also Micro-aneurysms
    berry 134
    cerebral, hypertension and 147–9
    'false' 147
Angiomas, cerebral 192
Antibodies 44
    'cell-bound' 12, 14
    effector role in immunological reactions 19